The Physiology of
NEMATODES

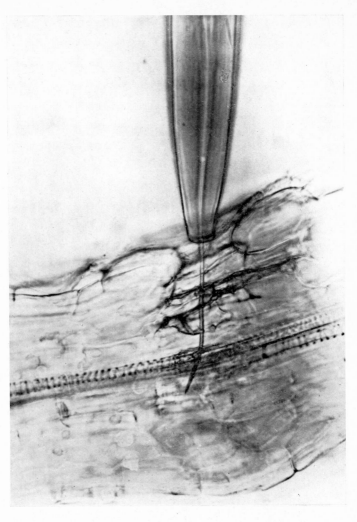

(*Frontispiece.*)   Larva of a plant parasitic nematode feeding
(by means of its stylet) upon the contents of the vascular
tissue in a plant root.   (Pitcher and Posnette. *Nematologica*,
1963.)

The Physiology of
NEMATODES

by

D. L. LEE

Molteno Institute of Biology and Parasitology
University of Cambridge
(Fellow of Christ's College Cambridge)

W. H. FREEMAN AND COMPANY
SAN FRANCISCO

First published .   .   . 1965

© 1965, D. L. Lee

Printed and published in Great Britain
by Oliver and Boyd Ltd., Edinburgh

Preface

Nematodes are usually neglected in undergraduate courses and the student is often left with the erroneous impression that most nematodes are parasitic in animals. Moreover, most of the courses are limited to a description of the morphology of *Ascaris lumbricoides* and to the life histories of a few economically important animal parasitic species and there is usually little or no mention of the physiology, biochemistry or behaviour of nematodes.

It is the aim of this book to present current knowledge of the physiology of nematodes (free-living as well as plant and animal parasitic species) in a form which will be useful to undergraduates and to University teachers.

The work which follows is based on many publications. Unfortunately it has not been possible to include references to all these, consequently most references are to books, reviews and recent papers and these will introduce the reader to the earlier literature. For this reason, reference to an author in the text does not imply that this author was the first to have made the observation in question.

I am deeply indebted to Mr F. G. W. Jones for much useful discussion and constructive criticism of most of the manuscript. I also wish to thank Mr R. G. Bruce, Dr D. W. T. Crompton and Dr P. Tate who read and commented on the manuscript at various stages of preparation, and Mr A. O. Anya, Dr C. Ellenby, Dr A. J. Munro, Prof. C. P. Read and Prof. W. P. Rogers for their comments on certain chapters. I am grateful to Miss S. Reeve who typed the manuscript and to Mrs R. Brand and Mr A. Page who helped with the illustrations.

I am also indebted to the following for permission to reproduce illustrations: Academic Press Inc. for figures 11, 12 and 21 from *The Nature of Parasitism* by W. P. Rogers, and figure 22 from *Experimental Parasitology*; Cambridge University Press for figures 4, 11, 12, 18, 19, 20, 29, 30, 33 from *Parasitology*, and figures 17, 26, 42, 43, 44, 45 from *Annals of Applied Biology*; Dr B. G. Chitwood for figure 7 and part of figure 9 from *An introduction to Nematology*; the Company of Biologists Limited for figure 37 from *Journal of Experimental Biology*; Macmillan & Co. Ltd. for figure 34 from *Nature, London*; E. J. Brill, Ltd., and East Malling Research Station for the plate from *Nematologica*; Rutgers University Press for figure 27 from *Host influence on parasite physiology*; The University of North Carolina Press for figures 3, 15, 16 and 17 from *Nematology. Fundamentals and recent advances with emphasis on plant parasitic and soil forms*; the Controller of Her Majesty's Stationery Office for figure 32 and part of figure 9 from *Plant Nematology*; The American Society of Tropical Medicine and Hygiene for figure 31 from the *American Journal of Tropical Medicine and Hygiene*; New Jersey Agricultural Experiment Station, Rutgers, for figure 39; *Journal of the Washington Academy of Sciences* for figure 35; *The Journal of Morphology* for figure 23; Akademische Verlagsgesellschaft Geest & Portig for figure 36 from *Zoologischer Anzeiger*; Veb Gustav Fischer Verlag for figures 38, 40 and 41 from *Zoologische Jahrbücher*; *Revue de pathologie comparée et d'hygiène générale* for figure 24. Acknowledgements are made to authors by mentioning their names in the legend.

This book contains some unpublished results of work supported by grant AI-04725-03 from the United States Public Health Service.

D. L. LEE

Molteno Institute of Biology and Parasitology,
Cambridge.

Contents

I: Introduction

There is a general belief amongst biologists that nematodes are mainly parasitic. This is because *Ascaris lumbricoides* is used as the 'typical' nematode in school curricula and because nematodes of medical, veterinary and agricultural importance are emphasized in later courses. Most nematodes are free-living: enormous numbers live in marine and fresh water mud and in the soil. Nematodes are widely distributed and are found in almost every type of environment. 'They occur in arid deserts and at the bottom of lakes and rivers, in the waters of hot springs and in the polar seas where the temperature is constantly below the freezing point of fresh water. They were thawed out alive from Antarctic ice in the Far South by members of the Shackleton Expedition. They occur at enormous depths in Alpine lakes and in the ocean. As parasites of fishes they traverse the seas; as parasites of birds they float across continents and over high mountain ranges.'[28]

The typical nematode is spindle-shaped, unsegmented and bilaterally symmetrical (figs. 1, 2). The adoption of a parasitic life has done little to alter the general shape of nematodes and, as Harris and Crofton[61] state, 'the elementary student may be forgiven at times for thinking that there is only one nematode but that the model comes in different sizes and with a great variety of life histories.' Cilia, flagella, respiratory and circulatory organs are absent and the excretory organs are unlike those of other invertebrates.

The outer body wall is composed of a cuticle, hypodermis and a layer of longitudinal muscle; the alimentary tract consists of a

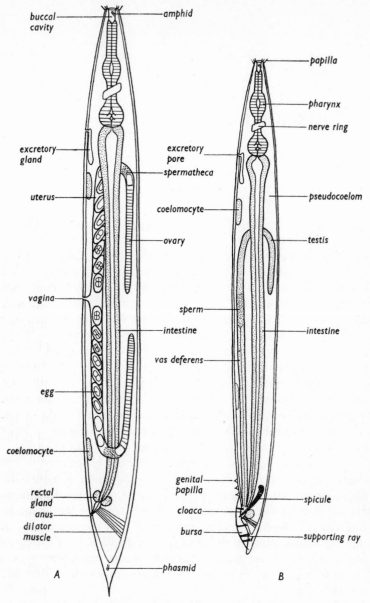

FIG. 1. General morphology of a nematode (hypothetical).
A. Female; B. Male. Lateral view.

terminal mouth, buccal cavity, pharynx, intestine, rectum and subterminal anus (cloaca in the male). The body cavity, in which lie the gonads, is a pseudocoelom (figs. 1, 2, 3, 4).

The sexes are almost always separate. Parthenogenetic females occur in some species and in others hermaphrodites are found, the gonad first producing sperm and later producing eggs.

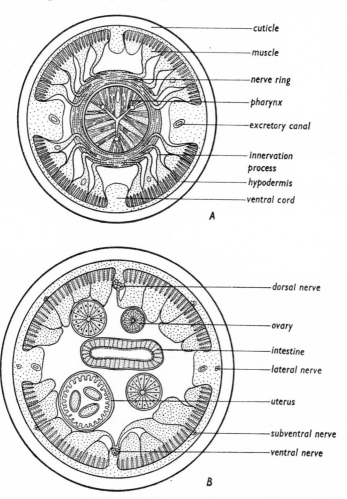

FIG. 2.   Transverse sections through *A*, the pharyngeal region and *B*, the middle region of a nematode.

The males are usually smaller than the females and possess copulatory aids such as genital papillae and spicules.

Nematodes undergo four moults from egg to adult worm. The larva, or juvenile as it should more correctly be called, is similar to the adult nematode except for the differences in size and in mouth parts, and in the lack of gonads and copulatory structures. At each moult the cuticle covering the body and lining the pharynx and rectum is shed, the new cuticle being formed beneath the old. Some nematodes moult once or twice within the egg before hatching; others, at a certain stage in the life cycle, retain the cuticle of the previous stage and this acts as a protective covering against adverse conditions. After the fourth moult the nematode emerges as the immature adult and development of the gonads occurs. The adult nematode may continue to grow in size without further moults (e.g. *Ascaris*, *Heterodera* females and *Meloidogyne* females).

The Organization of the Nematode Body

Cuticle

The cuticle of nematodes plays an important part in the physiology of the animal and will therefore be dealt with in some detail. It is a complex structure which varies from one genus to another and from larval stage to adult. Most work has been done on the cuticle of *Ascaris lumbricoides* which has three main regions made up of nine separate layers[7, 26] (figs. 3, 4). The surface of the cuticle is covered by a thin layer, less than 1000 Å in thickness, which is thought to consist of lipid. The main outer region is the cortex which is divided into outer and inner cortical layers. The amino acid composition of the cortex has led to the suggestion that this part of the cuticle is formed of keratin,[52] although X-ray diffraction studies indicate the presence of collagen. Polyphenol oxidase is present in this region and there is some evidence that it consists of quinone-tanned protein. Thus the cortex of the cuticle of *Ascaris* may be keratinized or subject to polyphenol-quinone tanning, or both.[48]

The middle region of the cuticle of *Ascaris*, called the matrix layer, is subdivided into the outer fibrillar, the thick homogeneous and (fig. 4) the thin boundary layers. The fibrillar layer contains

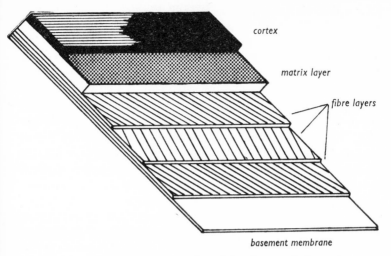

FIG. 3.   Schematic representation of the three main layers in the cuticle of *Ascaris* (after Fairbairn[48]).

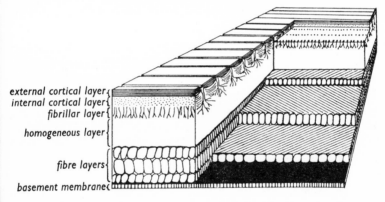

FIG. 4.   Diagram showing transverse, longitudinal and tangential sections of the cuticle of *Ascaris* (after Bird and Deutsch[7]).

distinct branching pore canals which extend into the external cortical layer.   These pore canals are hollow but are normally filled by a substance which is removed by peptic digestion, suggesting that it is rich in aromatic amino acids.[7]   The homogeneous layer is made up of albumen-like proteins of low molecular weight and also contains fibrous proteins which resemble fibroin or elastin.[26, 91] Small amounts of carbohydrate and lipid are also found. Esterase enzymes are present in this layer[82] and this, together with the polyphenol oxidase in the cortex, demonstrates that the cuticle is metabolically active and not an inert covering. These enzymes, which probably originate in the hypodermis, may take part in the growth of the cuticle of *Ascaris*. This species, after the final moult, is only a few millimetres long whereas the mature adult is approximately 20 cm. long. Unlike arthropods, where the hardened exoskeleton has to be discarded to enable increase in size to take place, this increase in size involves a continuous growth of the cuticle.

The innermost of the three main regions is the fibre layer which is composed of outer, middle and inner fibre layers and the basal lamella. Staining properties and X-ray diffraction studies suggest that these layers consist of collagen. The presence of hydroxy-proline also tends to confirm the presence of collagen.[48] These fibre layers cross each other at an angle of 135 degrees and enclose between them a system of minute parallelograms (figs. 3, 37). The angle between any side of the parallelogram and the longitudinal axis of the worm is about 75 degrees.[61]

The cuticles of other nematodes have not been studied in as much detail as that of *Ascaris* but, while there are usually differ-ences in the actual number of layers present in the cuticle, adults of the larger nematodes possess the three main regions of cortex, matrix and fibre layers. In smaller nematodes, however, the fibre layer is reduced or absent. The cuticle of *Oxyuris equi* consists of eight separate layers and the two inner fibre layers cross each other at an angle of 120 degrees and make an angle of 60 degrees with the long axis of the worm. The cuticle of *Strongylus equi* is similar to that of *Ascaris* but the pore canals of the fibrillar layer are hollow and pass from the homogeneous layer to just beneath the grooves of the external cortical layer. Pore canals are not found in the cuticle of *Oxyuris*.[7]

The cuticle of *Nippostrongylus brasiliensis* consists of the three main layers of cortex, matrix and fibre layers but the matrix layer is a fluid-filled layer which contains haemoglobin and some enzymes together with structures (called supporting struts) which support longitudinal ridges in the cuticle. There are two fibre layers (fig. 5). Numerous collagen fibres join the fibre layer, the supporting struts and the cortex.[83] The chemical properties of other nematode cuticles are fundamentally the same as those of *Ascaris* cuticle.[26]

Less work has been done on the structure of the larval cuticle. The cuticle of the larval stages of *Metastrongylus* and *Dictyocaulus*

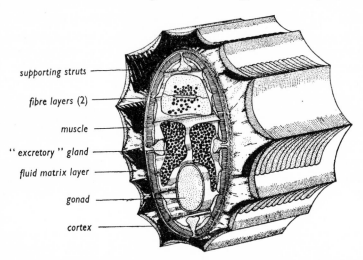

supporting struts ——

fibre layers (2) ——

muscle ——

" excretory " gland ——

fluid matrix layer ——

gonad ——

cortex ——

FIG. 5. A stereograph of a section taken from the middle region of *Nippostrongylus brasiliensis* to show the structure of the cuticle (Lee[83]).

is reported to be structureless[91] but there is evidence that tanned protein is present. The cuticles of these larvae are not attacked by papain, a plant enzyme which rapidly digests the cuticle of *Ascaris*. The cast cuticle of the fourth-stage larva of *Nippostrongylus* contains tyrosine, hydroxyproline, carbohydrate and lipid and the physical properties of the cuticle are those of an unusual collagen.[138] These analyses were on the cast cuticle and there may have been absorption of part of the cuticle during the moulting process, as occurs in arthropods. The cuticle of the third-stage larva of *Nippostrongylus* appears to have a thin outer layer and a thick fluid layer containing widely separated fibres, but no fibre layer such as occurs in the adult.[83] The cuticle of

B—P.N.

Trichinella spiralis larva is made up of two definite layers separated by a membrane. The inner layer has an array of very fine fibrils about 40 Å thick, arranged parallel to the circumference of the larva. There is also a thin external membrane present.[3]

The hypodermis

The hypodermis lies beneath the cuticle and may be cellular or a syncytium. It projects into the body cavity along the mid-dorsal, mid-ventral and the lateral lines to form four ridges or cords (figs. 2, 5). The lateral hypodermal cords are the largest and contain the excretory canals when these are present (fig. 2). The nuclei of the hypodermis are found only in the cords. The hypodermis contains large amounts of reserve materials (fat, glycogen) and, in some animal parasitic nematodes, also haemoglobin.

The nervous system

The nervous system consists of a circumpharyngeal ring (figs. 1, 2) with associated ganglia, large ventral and smaller lateral and dorsal nerves which run posteriorly from the nerve ring along the cords (fig. 2), and several nerves which extend to the anterior end where they innervate the sense organs around the mouth. There is a pharyngeal-sympathetic nervous system consisting of three nerves which run along the length of the pharynx and are united to each other by a few commissures.

The muscles

The muscles of the body wall of nematodes are unique. They are all spindle-shaped longitudinal muscles and the cells are divided into a contractile and a non-contractile portion (figs. 2, 6). The contractile portion of the cell consists of supporting elements between which lie the myofibrils (fig. 6). The non-contractile part of the cell contains the nucleus, mitochondria, glycogen and fat. The innervation of the muscles is most unusual: processes from the non-contractile part of the cell (innervation processes) pass from the muscle to the longitudinal nerves or the nerve ring (fig. 2). The muscles are attached to the cuticle by fibres which

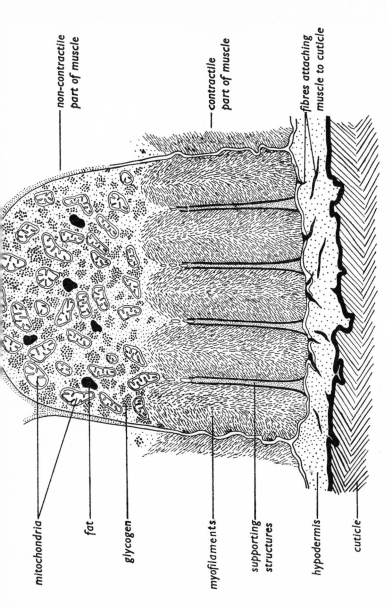

non-contractile part of muscle

contractile part of muscle

fibres attaching muscle to cuticle

mitochondria

fat

glycogen

myofilaments

supporting structures

hypodermis

cuticle

FIG. 6. Transverse section through a muscle of the body wall of *Nippostrongylus brasiliensis* towards the end of the muscle. Drawn from several electron micrographs.

run from the contractile part of the cell, through the basement membrane into the fibre layers (fig. 6).

The pseudocoelom

The body cavity of nematodes is usually called the pseudocoelom and is filled with fluid. It also contains large fixed cells, the coelomocytes. The pseudocoelom is an important structure as the fluid in it bathes all the internal organs and also forms the hydrostatic skeleton of the nematode. Nutrients from the intestine and oxygen and ions from the environment must cross the pseudocoelom to reach the gonads and the body wall. It is not surprising, therefore, to find that the pseudocoelomic fluid is a very complex solution, that of *Ascaris* for example contains a variety of proteins, fats, carbohydrates, enzymes, nitrogenous compounds and inorganic ions.[47, 48]

The pseudocoelomic fluid is always under pressure, owing to the tonicity of the body wall musculature, and this is of great significance in locomotion, feeding and excretion.[61]

Coelomocytes

The coelomocytes (also called stellate cells, athrocytes, phagocytic or giant cells) are ovoid or many-branched bodies situated in the pseudocoelom (fig. 1) and are usually two, four or six in number. Various functions have been attributed to these structures. They have been thought to be phagocytic, for in *Ascaris* the coelomocytes engulf bacteria injected into the body cavity, although ink particles are not phagocytized.[26] Another function ascribed to these structures is the storage of insoluble waste products but there is little evidence in support of this. The uptake and concentration of methylene blue and neutral red by these cells does not necessarily indicate an excretory function for, as in other animals, many types of cell take up these stains. They may be absorptive in function and ' purify ' the body fluid in some way;[26] on the other hand they may be secretory and release enzymes into the body fluid. It has also been suggested that they are an oxidative centre for the nematode.[66] Early workers believed that they were in continuity with the excretory system but this has been shown not to be so. Their physiological rôle in nematodes thus remains an unsolved problem.

Alimentary system

There is a mouth, surrounded by lips, a buccal cavity of varying shape, a muscular and glandular pharynx (which is a syncytium and has a triradiate lumen), an intestine, rectum and anus (figs. 1, 2). The pharynx, which is a powerful pumping organ, contains three glands which open into the lumen at varying positions along the length of the pharynx. The intestine is lined by microvilli and appears to be both secretory and absorptive in function. The alimentary system is dealt with in greater detail in chapter 2 (Feeding and Digestion).

Excretory system

The excretory system of nematodes was originally assigned this function on purely morphological grounds. It is a very varied system (fig. 7) and in some groups appears to be completely absent.[26]

The so-called excretory system is of two types, glandular and tubular. The glandular system is found in many free-living nematodes. It consists of a ventral gland cell situated in the region of the base of the pharynx and usually has a terminal ampulla which opens to the exterior on the ventral surface by means of a pore (figs. 1, 7*H*). The tubular system varies from a simple H-type system, as seen in the oxyurids (fig. 7*D*) in which a lateral canal runs along the length of the nematode in each lateral line and connects ventrally in the region of the base of the pharynx with the other canal and the excretory pore, to a reduced form in which one lateral canal completely disappears (fig. 7*C*, *G*). In *Ascaris* and some other nematodes the H-type has become modified into an inverted U-type system, the anterior limbs of the H being absent (fig. 7*E*, *F*). Some nematodes possess both lateral canals and ventral glands (fig. 7*A*, *B*). The lateral canals are believed to be intracellular as the whole system has only one nucleus. There is a complete absence of flame cells in this system, but in certain nematodes pulsations of various parts of the system have been observed.

Reproductive system

The sexes are usually separate, the males frequently being smaller than the females. Males have one or two testes which

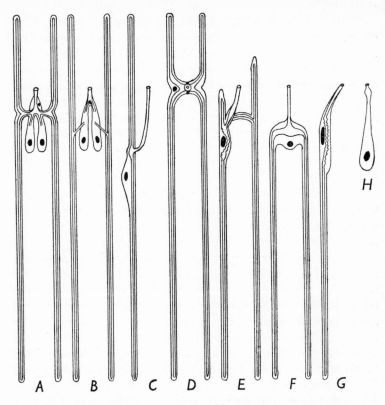

FIG. 7. Representative types of excretory system found in
nematodes. *A*. Rhabditoid type, an H-shaped system with
two subventral gland cells, the lateral canals are situated in
the lateral cords; this type is found in *Rhabditis*, *Rhabdias*
and the Strongylina in general. *B*. Variant of *A*, found in
Oesophagostomum. *C*. Tylenchoid type, an asymmetric
system with the lateral canals and gland cell confined to one
cord. *D*. Oxyuroid type, an H-shaped system without sub-
ventral glands, with a greatly shortened terminal duct. *E*.
Ascaroid type, a shortened H-type nearly an inverted U in
form. *F*. Cephaloboid type of excretory system. *G*.
Anisakid type, a reduced form related to *E* in which one
lateral canal has completely disappeared. *H*. Single ventral
cell type, present in Chromadorina, Monhysterina and
Enoploidea (after Chitwood and Chitwood[26]).

open into a seminal vesicle, posterior to which is the vas deferens. This latter structure may be divided into glandular and ejaculatory parts. The vas deferens opens into a cloaca. Many male nematodes possess copulatory spicules (fig. 1*B*) which lie in pouches formed from the cloaca and the area of the nematode around the cloaca may be expanded to form a copulatory bursa. The spicules consist of sclerotized cuticle with a cytoplasmic core and in some cases (*Ascaris*) a nerve runs to a sense organ at the tip.[82] The sperm of nematodes are amoeboid.

The females usually have one or two ovaries which open into oviducts and the uterus. The sperm are stored at the ovarian end of the uterus (fig. 1). Muscle fibres cover the tubes of the female gonads and bring about movement of the contents. The uteri end in an ovijector which is very muscular and which, together with the high internal pressure of the body fluid, serves to expel the eggs through the vagina.

In some species of nematodes the sex ratio is density dependent and is possibly controlled by nutritional factors. Thus, in *Mermis* the eggs develop as females or males depending on the number which are eaten by the grasshopper, a preponderance of males occurring in heavily infected grasshoppers. Similarly, sex determination in *Heterodera rostochiensis* and *Meloidogyne* apparently depends on the abundance of food, for the proportion of the sexes varies with the intensity of infestation of the host, more males appearing in large populations. Furthermore, female larvae of *Meloidogyne* can undergo sex reversal, if the nutritive conditions are altered, and develop into adult males with two testes (normal males have one testis).[152]

Distribution

Nematodes are found in every ecological environment with the exception of the aerial and pelagic habitats. Although they are essentially aquatic animals some species are anabiotic (cryptobiotic) and can survive for long periods in a dry state, becoming active when water becomes available.

Nematodes are probably the most abundant of the small metazoa on the ocean bed, several millions being found per square metre in the top few centimetres. Nematodes are also very abundant in fresh water muds and in soil. Some species have peculiar habitats, for example, malt vinegar, felt beer mats, hot springs and the pitchers of the pitcher plant.

Nematodes feed on living plants, either as ecto- or endo-parasites, and are one of the important groups of invertebrate parasites of animals (the other groups being the protozoa, the platyhelminthes and the arthropods). Because they have filled so many ecological niches and because of their great numbers, nematodes must be regarded as one of the most successful groups of animals. Nevertheless, surprisingly little is known about the physiology, biochemistry, behaviour and anatomy of nematodes. Most work has been done on a few large nematodes parasitic in animals and little is known about the physiology of free-living nematodes or of most of the small parasitic ones.

2 : Feeding and Digestion

Nematodes feed upon a wide variety of foods but most individual species are restricted to one type of food. There are saprophagous, microbivorous, phytophagous, predatory and parasitic species. Marked differences exist in the structure and physiology of the digestive system of nematodes. These differences are, however, not nearly so pronounced as in insects, which are another group of invertebrates with extremely varied feeding habits.

The Alimentary Canal

As far as is known, all nematodes take their food into an alimentary canal which is fundamentally the same in all groups. It is a straight tube from terminal mouth to sub-terminal anus and is divided into three parts:—

1. The **stomodaeum** which includes the mouth, lips, buccal cavity and pharynx (oesophagus);
2. the **intestine**;
3. the **proctodaeum** consisting of the rectum and anus in females and the cloaca with its associated structures in the males. The stomodaeum and the proctodaeum are lined with cuticle which is shed at each moult.

The high turgor pressure in the fluid-filled pseudocoelom of nematodes has an important bearing on the structural organization of the alimentary system. The intestine is normally collapsed, except when full of food, and the pharynx has to produce an even

15

higher pressure to force food into it. The pharynx and anus are provided with devices which are closed by the pressure of the body fluid and opened by dilator muscles.[61]

Stomodaeum

The greatest structural diversity in the alimentary system occurs in the stomodaeum and this is associated with the varied feeding habits of nematodes. In free-living nematodes there are usually six lips surrounding the mouth but in some nematodes they have become reduced in number by fusion (*Ascaris* has three) and this is usually associated with the parasitic habit. In other nematodes the lips form elaborate processes possessing many spines (*Acrobeles*), hooks or a type of sucker (fig. 8).

The buccal cavity is that part of the alimentary tract between the mouth and the pharynx and varies greatly in shape (fig. 8). It may be cup-shaped and contain teeth or cutting plates, as in some predatory nematodes (*Mononchus*) (fig. 8*E*) and some parasitic species which feed upon the tissues of other animals (*Syngamus*; *Ancylostoma*) (figs. 8*G*, *H*). It may be merely a cylindrical narrow tube associated with bacterial or fluid feeding as in *Rhabditis* (fig. 8*A*) and *Ascaris*, or a stylet may be present and used to pierce plant cells, animal prey or tissues so as to extract the contents[26] (*Actinolaimus*, *Dorylaimus*, *Tylenchus*, *Heterodera*, *Trichuris*) (figs. 8*B*, *C*, *D*). In the Tylenchida the stylet is a hollow cuticular structure which develops in the wall of the buccal cavity whereas in the Dorylaimida the stylet develops in the pharynx. These stylets can be extruded by muscles.

The buccal cavity opens into the pharynx which is a syncytial, muscular and glandular pumping organ with a tri-radiate lumen. In some nematodes both the median and posterior parts of the pharynx are swollen to form muscular bulbs, in others only one bulb occurs, while some have none (fig. 9). The lumen of the pharynx is lined with cuticle and in those nematodes which possess a posterior pharyngeal bulb the cuticular lining of the bulb is corrugated and forms three plates or flaps which sometimes act as a grinding mechanism[38] (figs. 9, 10). Feeding in microbivorous species, which have a pharynx similar to *Rhabditis*, is described in figure 10. In nematodes without this bulb the food collects at the posterior end and passes into the intestine when the

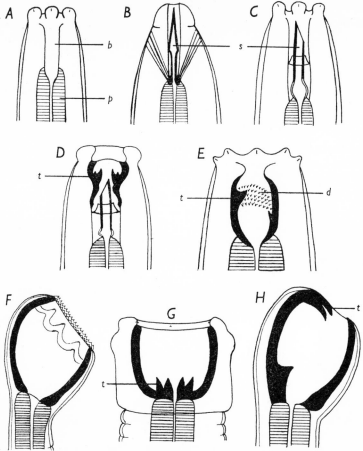

FIG. 8. Diagrams of some nematode heads showing variations in the structure of the buccal cavity. *A. Rhabditis* (microbivorous); *B. Tylenchus* (plant parasite); *C. Dorylaimus* (some are phytophagous and some are predators); *D. Actinolaimus* (predator); *E. Mononchus* (predator); *F. Chabertia* (animal parasite); *G. Syngamus* (animal parasite); *H. Ancylostoma* (animal parasite).

b, buccal cavity; *d*, denticles; *p*, pharynx; *s*, stylet; *t*, tooth.

pharyngeal-intestinal valve opens. Radial muscles bring about dilation of the lumen of the pharynx from the anterior to the posterior end thus producing a sucking action. Valves which prevent the regurgitation of food are present in the pharynx.

There are three glands in the pharynx, one dorsal and two sub-ventral.   In some species the dorsal gland opens into the buccal cavity while in others it opens into the anterior end of the pharynx or more posteriorly.   The two ventral glands open into the posterior part of the pharynx in some species but in others they open anteriorly (fig. 9).

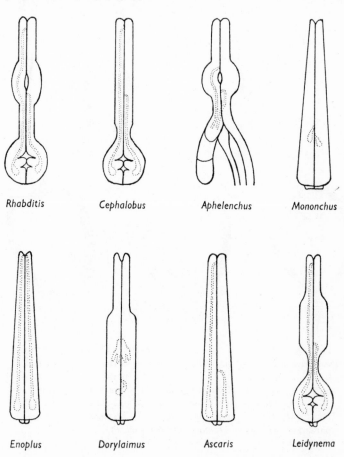

Rhabditis   *Cephalobus*   *Aphelenchus*   *Mononchus*

Enoplus   *Dorylaimus*   *Ascaris*   *Leidynema*

Fig. 9.   Diagrams of some nematode pharynges.   The positions of the pharyngeal glands and ducts are shown by broken lines.   (Partly after Jones, 1959.   *Plant Nematology*, H.M.S.O.;   partly after Chitwood and Chitwood[26].)

Intestine

The intestine is divided into three regions, the anterior (ventricular) region, the mid-region and the posterior (pre-rectal) region. These regions differ from each other in the shape of the lumen, the height and the contents of the cells, and possibly in their function. The wall of the intestine consists of a single layer of cells and there is usually a basement membrane covering the external surface of these cells. When present this membrane may take the form of a thick, collagenous basal lamella (*Ascaris*), a muscle network (*Leidynema*; *Oxyuris*) or a simple membrane (*Nippostrongylus*). The internal surface of the cells is covered with numerous microvilli[72, 160] which previously were referred to as the bacillary layer or striated border before their true nature was determined. Similar structures are present on the intestinal cells of many animals and are regarded as a device for increasing the absorptive area of the intestine. The intestinal cells are believed to be both secretory and absorptive in function but there is some evidence that the anterior region is mainly secretory and the mid- and posterior regions mainly absorptive.[22, 44]

The contents of the intestine move about as a result of the locomotory activities of the nematode and the ingestion of more food, together with peristaltic movements of the intestine in some species.

In most nematodes the intesto-rectal valve opens periodically and the contents of the pre-rectal region are ejected with considerable force from the rectum and anus, because of the high internal pressure of the body fluid. In some, however, the muscle network covering the intestine helps to force the waste material out.

Proctodaeum

The rectum is a short, cuticular-lined structure and in many nematodes rectal glands of unknown function open into it. In the male the alimentary system and the reproductive system open into a cloaca which contains the spicules and accessory copulatory structures when present. In some nematodes the intestine ends blindly and there is no anus (*Mermis*).

Feeding

Microbivorous and saprophagous

Included in this group are nematodes which feed on either bacteria or the products of bacterial decay. Bacterial and fluid feeding nematodes usually have a simple alimentary system with little modification of the mouthparts (*Rhabditis*) (fig. 8*A*), but in some there are small teeth at the base of the buccal cavity and in others the oral region has become elaborated into processes (prolabae) possessing many spines (*Acrobeles*). Many species of *Rhabditis* inhabit the soil and feed on bacteria or decaying organic matter which they draw into the alimentary tract by means of the sucking action of the pharynx (fig. 10). *Turbatrix aceti*, the vinegar worm, feeds in a similar manner upon bacteria and yeasts in malt vinegar. *Acrobeles* and related genera, which have elaborate prolabae around the mouth, also feed upon bacteria but the prolabae, with their associated bristles and spines, are believed to act as a sieve to restrict the size of the particles which the nematode ingests.

The pharyngeal glands produce a secretion, thought to be digestive, which is mixed with the food during its passage along the pharynx and in some nematodes the plates in the pharyngeal bulb may grind some of the particles before they are passed back into the intestine.[38]

The free-living stages of many nematodes, which as adults are parasitic in animals, are microbivorous. The nature of the food and the feeding mechanisms are essentially the same as in free-living *Rhabditis*; for example, the first- and second-stage larvae of *Ancylostoma* and of *Nippostrongylus* feed upon bacteria in the faeces of the host. Third-stage larvae are the infective stage and do not feed but exist on their food reserves.

Phytophagous

The phytophagous nematodes feed upon fungi, diatoms, algae and the cells of higher plants. They usually have a stylet with which they pierce the plant cell (frontispiece) (fig. 8*B*, *C*) and withdraw the cell contents.

Many of the stylet-bearing nematodes attack living cells of

higher plants either as ectoparasites (when they feed externally upon the cells or vascular tissue of the plants in a somewhat similar manner to aphids) (frontispiece) or as endoparasites (when they migrate into and feed upon the internal tissues of the plant).

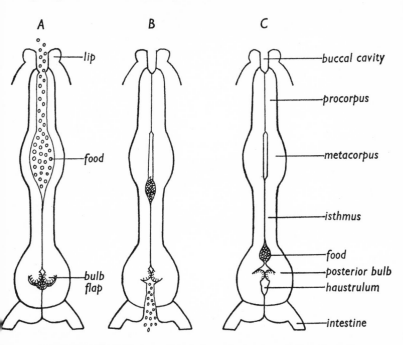

FIG. 10.   Diagrams to show the structure and function of the *Rhabditis*-type pharynx during feeding. Food particles, small enough to pass through the buccal cavity, are drawn into the lumen of the metacorpus by sudden dilation of the pro- and metacorpus (*A*).   Closure of the lumen of the pharynx in these regions expels excess water (*B*) and the mass of food particles is passed backwards along the isthmus by a wave of contraction of the radial muscles of the isthmus (*B, C*). Food is drawn between the bulb flaps of the posterior bulb by dilation of the haustrulum, which inverts the bulb flaps (*A*), and is passed to the intestine by closure of the haustrulum and by dilation, followed by closure of the pharyngeal-intestinal valve (*B*). The bulb flaps contribute to the closure of the valve in the posterior bulb and also crush food particles when they invert (*A*). (Diagrams from a description given by Doncaster[38, 39].)

When feeding the nematode pierces the cell wall with the stylet and then sucks out the contents, usually through the hollow stylet (*Aphelenchoides, Ditylenchus, Heterodera, Xiphinema*).

When a plant parasitic nematode reaches the roots of the host plant it probes cells without actually puncturing them. These probing movements probably represent a search for a suitable spot to attack. *Heterodera*, for example, usually attacks at a weak point, such as the junction of a lateral root with the main root. When penetration of the plant cell begins the body of the nematode arches, bringing the stylet down at right angles to the cell surface. This allows maximum thrust to occur. The stylet is then repeatedly thrust against the cell wall until the cell is punctured. During penetration the nematode must be suitably anchored by external resistances, such as surface tension forces or surrounding soil particles, otherwise the nematode will be pushed away from the cell.[152] The nematode may also anchor itself to the roots of the plant by means of suction between fused lips and the hydrophobic surface of the cell.

Following penetration saliva is discharged through the stylet into the cell. Cellulase, chitinase, pectinase, amylase and other hydrolytic enzymes are present in the saliva of many phytophagous and plant parasitic nematodes (table I) and these will soften the cell wall, making it easier to puncture and also bring about extracorporeal digestion of the cell contents. The contents of the cell are then withdrawn by means of the inserted stylet and the sucking action of the pharynx.

Nematodes, which enter the tissues of the plant, first penetrate the epidermis or enter through lenticels or stomata. They then migrate through the tissues, usually by pushing and digesting their way between cells. Some species migrate through the tissues, feeding on different cells as they move along, while others are responsible for the formation of giant cells upon which the nematode feeds. The giant cells are formed by a combination of turgor pressure and the dissolution of cell walls resulting in the merging of protoplasm from the cells to form a syncytium. The increase in turgor pressure probably occurs because of hydrolysis of starch in the cells by amylase in the saliva of the nematode. *Meloidogyne* larvae, after invasion of the root, migrate both inter- and intracellularly to the site where feeding and the subsequent swelling and development occurs. Feeding by the nematode results in the formation of giant cells. Females of *Heterodera*, once inside the root, also remain feeding in one position. The body of the nematode swells, rupturing the cortex of the root, but the head region remains anchored in the tissues, apparently by a sticky secretion produced by the nematode. The head region has a certain amount of mobility and the nematode feeds upon the giant cells surrounding it. Most species of *Ditylenchus* live intercellularly and are responsible for loosening and necrosis of plant tissues. They feed on the intercellular pectin and the cell contents.[152]

One fungal feeding nematode injects bacteria from the alimentary tract into the mycelium of mushrooms; the bacteria rapidly multiply in the fungus and the nematode feeds upon the bacteria and the products of decay they produce.

Many nematodes without stylets feed upon diatoms and blue-green algae.

Carnivorous

Some stylet-bearing nematodes prey upon protozoa, nematodes, rotifers, tardigrades, small oligochaetes or even upon the eggs of other nematodes. The method of feeding is very similar to that in the phytophagous nematodes which have stylets; in fact many of these predatory stylet-bearing nematodes belong to genera in which the majority of species are phytophagous. In the genus *Aphelenchoides* some species are fungal feeders, while others are plant parasites, predators or insect parasites. Similarly, most members of the genus *Dorylaimus* are microbivorous or phytophagous but some species are carnivorous and feed upon other nematodes. Some nematodes are omnivorous and can feed upon a selection of different foods such as algae, protozoan cysts, fungi and other nematodes.

When feeding the nematode must be suitably braced against external resistances, the lips are closely applied to the prey, the stylet is thrust into the prey and saliva, containing enzymes, is secreted into the wound. The digested contents are then sucked out through the stylet. Many of these predators may secrete a paralysant into the prey but puncturing the body wall of many small animals, especially nematodes, must cause reduction in body

C—P.N.

turgor pressure which will immobilize the prey and enable feeding to proceed without the prey struggling to escape. A species of *Nygolaimus*, which feeds upon small oligochaetes, has a stylet which lacks an opening. The stylet is used to pierce the worm, after which it is retracted and the contents are sucked out by the pharynx.[146]

Many predaceous nematodes do not have a stylet but possess powerful teeth or jaws and the buccal cavity is lined with small denticles (fig. 8E). These puncture, shred and bite the prey. Some nematodes possess both teeth (onchia) and a stylet (fig. 8D), the teeth holding the prey while the stylet is thrust into it (*Actinolaimus*). Many of the jawed predators swallow their prey whole if it is small enough, rupture the cuticle with their teeth and suck out the contents, or bite pieces off and swallow the pieces whole. *Mononchus* feeds on protozoa, nematodes, rotifers, tardigrades and small oligochaetes. Captured nematodes are often seen whole or partially digested in the intestine together with hard indigestible structures such as the stylets from nematodes or the chaetae from oligochaetes. One individual *Mononchus* killed 1332 nematodes in 12 weeks in one experiment.[79]

Parasites of animals

Several species of nematodes, belonging to genera which are normally microbivorous or phytophagous, are parasitic in insects. Species of *Tylenchus* and *Aphelenchoides* live in the tunnels made by bark beetles and are often transported on the bodies of the beetles; a few species, however, have become parasitic in the beetles, the larval stages living in the haemocoele or Malpighian tubules. Larvae of species of *Diplogaster* also parasitize the internal organs and haemocoele of beetles. *Neoaplectana* is a facultative parasite which lives in the intestine, and later invades the tissues, of certain insects.

Mermis, in its larval stages, is parasitic in insects and other invertebrates but the adult is free-living. The larvae possess a stylet with which they attack the viscera and fat body of the host. The intestine becomes a storage organ, the cells storing food materials upon which the free-living adult later exists.

Pre-larvae (microfilariae) of the filarial nematodes are transmitted by biting flies. The larvae feed upon the tissues (wing muscle, fat body) of the fly, probably making use of histolytic secretions as well as a small buccal tooth.

Leidynema is an oxyurid nematode which lives in the hind-gut of cockroaches where it feeds upon the contents of the hind-gut. It has a powerful sucking pharynx but no stylet or teeth and it never attacks the tissues of the host.[83]

Nematodes which live as parasites in vertebrates can be grouped by their methods of feeding:—[26, 79]

(*a*) Nematodes feeding on the contents of the alimentary canal of the host (*Ascaris, Oxyuris, Heterakis, Ascaridia*).

Typical nematodes of this group are *Ascaris lumbricoides*, which feeds on the intestinal contents of the pig and of man; and *Ascaridia galli*, which feeds on the intestinal contents of the domestic fowl.

Charcoal, barium and other materials have been fed to animals infected with *Ascaris* or related nematodes and these substances have been ingested by the nematodes. *Ascaridia* takes up radio-active phosphate when it is fed to the host but not when the radioactive phosphate is injected into the blood stream of the host (fig. 11), showing that this nematode feeds on the intestinal contents and not on the blood or tissues of the host.[122]

These nematodes feed by drawing semi-digested food, bacteria and other intestinal contents into the alimentary system by means of the powerful sucking pharynx. The food is then passed along the pharynx, where secretions from the pharyngeal glands mix with it, to the intestine where digestion is completed and absorption takes place. The lips are reduced in number in several species of this group and the buccal cavity is simple. They are essentially microbivorous and saprophagous feeders.

(*b*) Nematodes feeding on the mucosa of the alimentary or respiratory tract and attached by a large buccal cavity or capsule (*Ancylostoma, Necator, Uncinaria, Chabertia, Syngamus*) (fig. 8F, G, H).

Included in this group are the hookworms of economic importance, *Ancylostoma* and *Necator*. These species browse on the intestinal mucosa of the host and often cause severe bleeding. They possess a large buccal cavity (fig. 8H) with cutting plates or hooks at the entrance and/or sharp teeth at the base of the capsule. The nematode feeds by drawing a plug of tissue into the buccal cavity and abrading it with the cutting plates and teeth. Digestive enzymes and an anticoagulant are secreted from the pharyngeal glands. The abraded tissues are digested in the intestine. *Ancylostoma caninum* appears to seek out blood vessels from which

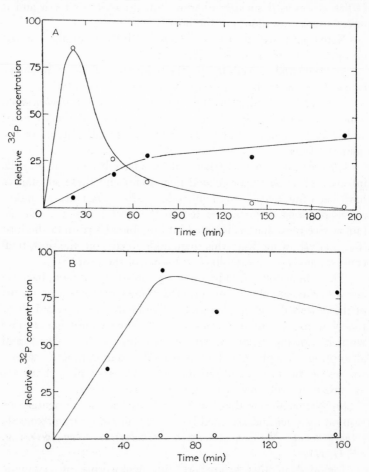

FIG. 11. The relative amounts of radioactive phosphorus found in chicken gut tissues (black dots) and *Ascaridia galli* females (circles) at several periods after dosing infected chickens (*A*) by mouth and (*B*) intravenously with disodium hydrogen phosphate containing [32]P (after Rogers and Lazarus, 1949. *Parasitology*[122]).

it extracts blood very rapidly (120 dilations and contractions of the pharynx a minute) and a female, which is 14–16 mm. in length, can extract 35 times its wet weight of blood in 24 hours.[115]   Blood is rapidly pumped into the intestine, but the nematode digests very little of it as large quantities of blood are passed out of the anus apparently unchanged.   Many nematodes, which have well developed buccal capsules but have no teeth or cutting plates (fig. 8F), feed by drawing a plug of tissue into the capsule and liquefying it by means of pharyngeal secretions.   They then ingest the fluid and semi-fluid material (*Chabertia*).

(*c*) Nematodes, without a large buccal capsule, which feed by penetration and histolysis of the tissues, or by puncturing the tissues (*Trichuris, Anisakis Dioctophyme, Nippostrongylus, Haemonchus*, infective larvae of *Ancylostoma* and *Nippostrongylus*).

An example of a nematode which feeds by penetration and histolysis of the host tissues is *Trichuris*, which buries its anterior end in a tunnel in the intestinal mucosa of the host and produces a histolytic secretion.   The liquefied tissues are then ingested by the nematode.   Both *Trichuris vulpis* and *Trichinella spiralis* adults possess a mouth stylet which, by repeated protrusion and retraction, is used to lacerate host tissues, small lymphatics and other vessels.   This releases cell contents and other fluids which, together with the products of histolysis, are ingested by active pumping of the pharynx.[26]

Dioctophyme renale lives in the kidney of dogs and has a large dorsal pharyngeal gland the secretion from which produces extensive cytolysis of the kidney.   The nematode then feeds on this semi-digested material.

The first- and second-stage larvae of *Ancylostoma* and of *Nippostrongylus* are microbivorous but the third-stage larva penetrates the skin of the host by means of digestive secretions and could thus ingest the food materials broken down by the enzymes.

Many nematodes have poorly developed buccal cavities but are able to draw blood from the tissues of the host.   *Haemonchus contortus* makes cutting movements with the tooth in its mouth and feeds on the released blood and tissue fluids.   An anticoagulant is secreted and injected into the wound.[135]   The adult nematode does not penetrate into the mucosa as does *Trichuris*. The fourth-stage larva of *Haemonchus* burrows into the mucosa of the abomasum, however, and causes the formation of a small blood clot underneath which it lies and feeds on blood.

Phosphate labelled with ^{32}P, when injected into rats infected with *Nippostrongylus*, rapidly appears in the nematode, showing that *Nippostrongylus* feeds on the tissues and tissue juices of the rat (fig. 12).[122]

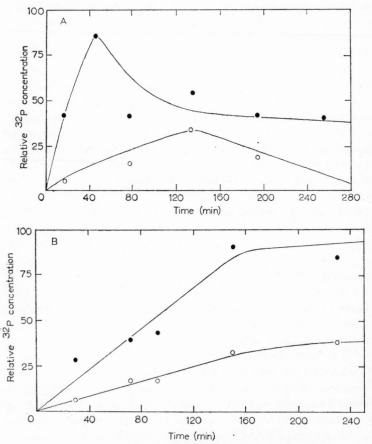

FIG. 12.   The relative amounts of radioactive phosphorus found in rat gut tissues (black dots) and *Nippostrongylus brasiliensis* (circles) at several periods after dosing infected rats (*A*) by mouth and (*B*) intramuscularly with disodium hydrogen phosphate containing ^{32}P (after Rogers and Lazarus, 1949.   *Parasitology*[122]).

In the Ascaroidea there is a relationship between the development of the pharyngeal glands and the method of feeding.
Ascarids which attack the mucosa and cause liquefaction of the
tissues have a well developed dorsal pharyngeal gland and it is
probable that the secretions from this gland are largely responsible
for the extracorporeal digestion of the tissues (*Anisakis*). *Ascaris*
and other ascarids which do not attack the host tissues but feed
on the contents of the intestine (group *a*) have a much less well
developed dorsal gland.[65]

(*d*) Nematodes living in body and tissue fluids and feeding on
the fluids (*Wuchereria, Dirofilaria*).

Adult filarial nematodes live in the lymph vessels, blood
vessels, coelomic spaces and in certain tissues of vertebrates.
They have no food-procuring organs other than the very small
mouth and narrow pharynx, which presumably take in liquid food
such as serum or lymph (*Wuchereria, Dirofilaria, Onchocerca*).

The larvae of many intestinal parasites travel through the
blood or lymphatic systems to the lungs on their migration to the
intestine and can thus feed on blood, serum or lymph (*Ascaris*
larvae, *Ancylostoma* larvae).

Digestive Enzymes

The digestive enzymes of nematodes appear to be related to the
type of food they ingest, but most nematodes are capable of
digesting carbohydrates, proteins and fats to a greater or lesser
extent (table I).

Enzymes in the pharynx

The glands in the pharynx of nematodes are thought to
secrete enzymes and certain other substances such as anti-
coagulants and paralysing substances, but most of the evidence
is circumstantial having been deduced from observing the effects
of the secretions on host tissues or on prey.

Predatory nematodes often carry out extracorporeal digestion
of their prey and the secretion from their pharyngeal glands
probably contains several digestive enzymes. *Diplogaster*, some
species of which are predators, secretes proteolytic enzymes
and a glycogen splitting enzyme.[96] *Actinolaimus* possesses a

cholinesterase and a less specific esterase around the base of the jaws and stylet and these enzymes may be used in feeding. The phytophagous nematode *Dorylaimus keilini* has no esterases around the base of the stylet.[83]

There is no information about the enzymes present in the pharynx of free-living microbivorous nematodes. *Turbatrix aceti* does not possess cellulase or chitinase.[150]

More is known about the enzymes in the pharynx from nematodes which feed on fungi and higher plants (table I). Several of these nematodes possess a cellulase, a pectinase and a chitinase which probably assist the nematode in puncturing the cell walls and in loosening the tissues of the plant.[150] *Ditylenchus dipsaci* secretes a pectinase into the onion and thus brings about breakdown of the intercellular pectin upon which it feeds. This results in the maceration of the tissues of the onion without releasing the contents of the cells which are toxic to the nematode. It secretes only 14 per cent. as much amylase as *D. destructor*, which feeds on potato starch.[95] Other fungus and plant-feeding nematodes secrete amylase, invertase and protease in their saliva.[53, 77, 95, 150]

The skin- and tissue-penetrating larvae of many animal parasitic nematodes carry out extracorporeal digestion of the tissues during entry and migration in the final host. They produce a ' spreading factor ' which has a similar action to hyaluronidase but this enzyme has been identified in a few species only (table I). Some skin-penetrating larvae bring about marked changes in the extracellular glycoprotein material of the skin and cause the basement membrane at the site of penetration to disappear. The enzyme is active on cartilage but there is little activity using native collagen as substrate.[84]

There is a lipase in secretions from the skin-penetrating stage of *Nippostrongylus*[147] and esterase is present in the pharyngeal glands and in the lumen of the pharynx at the anterior end. The adult nematode contains a large amount of esterase in the pharyngeal glands and the ducts of these glands.[83] These enzymes are believed to take part in extracorporeal digestion.

Several nematodes (*Ancylostoma*, *Haemonchus*) secrete an anti-coagulant which is thought to act by preventing the formation of thrombin from prothrombin.[135]

A proteolytic enzyme is present in the pharynx of *Ancylostoma caninum* and may play a part in extracorporeal digestion.[148] *Trichuris muris*, which burrows its pharyngeal region in the tissues of the host, has several enzymes which could be of use in carrying out extracorporeal digestion.   It is significant that the anterior end of the nematode has much higher proteolytic and esteratic activity than the posterior end.   None of the enzymes in *Trichuris* (table I) has been definitely implicated in penetration and digestion of the host tissues but some are almost certain to be involved.[99]

Many nematodes belonging to the Ascaroidea (*Anisakis*, *Contracaecum*) carry out extracorporeal digestion and the pharyngeal glands are believed to secrete enzymes which carry out liquefaction of the host tissues.[65]

Ascaris has several hydrolytic enzymes in the pharynx.[22]   There is an amylase, a maltase, a protease, peptidases hydrolyzing alanylglycine and leucylglycine, a lipase and an esterase.   The esterase has been located, histochemically, in the walls of the pharyngeal glands.[82]   There is no evidence that any of these enzymes are secreted by the pharynx.   Any pharyngeal secretion in *Ascaris* would almost certainly be mixed with the ingested food and would not be secreted to the exterior, as it is not a tissue feeder.

Secretion in the Intestine

Secretory activity in the intestine has been described in *Ascaris lumbricoides*.[22, 82]   Holocrine secretion occurs in the anterior region of the intestine, parts of the cell containing secretory material becoming detached from the intestine and then disintegrating in the lumen.   Merocrine secretion, in which droplets coalesce and pass through the cell border into the lumen, also occurs in *Ascaris*.[82]   The anterior part of the intestine contains many more secretory cells than the mid-intestine and posterior intestine.[22]

Enzymes in the Intestine

(a) Digestion of carbohydrates

Starch and glycogen.   Most nematodes are able to digest starch and glycogen and the amylases, which help to break down these substances, appear to be similar to those of other animals, although there are minor differences in pH optima and in activators and inhibitors of the enzymes.

Amylase is present in all of the phytophagous nematodes which have been investigated (table I), but the activity of the enzyme is related to the nature of the food. For example, *Ditylenchus destructor* which feeds on potato starch has a much more active amylase than *D. dipsaci* which feeds mainly on pectin.[95]

All animal parasitic nematodes have amylases in the intestine (table I).[22, 80, 99, 117] *Graphidium strigosum*, however, is only able to digest rice starch after the starch has been heated,[44] which would indicate that it possesses an amylase but was unable to digest the covering of the starch grain.

Cellulose, chitin and pectin. Enzymes capable of breaking down these substrates have been found only in nematodes which feed upon fungi or higher plants (table I).[150] The presence of cellulase, chitinase and pectinases is apparently related to the type of feeding of these nematodes for *Turbatrix aceti* (which feeds on bacteria and yeasts) has neither cellulase nor chitinase.[150] It is not known where these enzymes are sited, as homogenates of whole worms were used to detect them, but it is presumed that some of the enzymes are present in the pharyngeal glands.

Mucopolysaccharides. Most of the larvae of nematodes which penetrate the skin or tissues of the animal host possess enzymes which break down mucopolysaccharides. In many, the specific enzyme involved has not been identified and is referred to as a ' spreading factor '.[84] Some nematode larvae are known to possess hyaluronidase or mucopolysaccharidase activity;[85, 107] these enzymes are probably in the pharyngeal glands and are secreted in the saliva when penetration occurs.

Adult nematodes which live on the mucosa and tissues of other animals probably possess mucopolysaccharidase, although evidence of this or hyaluronidase activity has been detected in only a few species (table I).[99, 107]

Sugars. Very little work has been done on the sugar digesting enzymes of nematodes so that it is almost impossible to generalize on their nature and distribution. It will be seen from table I that maltase is present in the intestine of the two species that feed on

TABLE I

The hydrolytic enzymes present in the alimentary system of nematodes which have different methods of feeding and different foods

NEMATODE	Anticoagulant	Amylase	Maltase	Invertase	Lactase	Trehalase	Other glycosidases	'Mucopolysaccharidase'	Pectinase	Chitinase	Cellulase	Protease	Peptidase	'Collangenase'	Esterases	Lipase
Microbivorous																
Turbatrix aceti[150]	−	−	−	−	−	−	−		−	o	o	−	−	−	−	−
Diplogaster larvae[96]	−	+	−	−	−	−	−		−	−	−	+	−	−	−	−
Phytophagous																
Ditylenchus destructor[77,95]	−	+	−	+	−	−	−		+	+	+	+	−	−	−	−
D. dipsaci[95,150]	−	+	−	+	−	−	+		+	+	+	+	−	−	−	−
D. triformis[77,150]	−	+	−	−	−	−	+		+	−	+	o	−	−	−	−
Heterodera[53,93]	−	+	−	+	−	−	−		+	−	−	+	−	−	−	−
Pratylenchus[93]	−	−	−	−	−	−	−		−	−	−	+	−	−	−	−
Meloidogyne[95]	−	−	−	−	−	−	−		−	−	−	+	−	−	−	−
Animal parasitic tissue migrating larvae																
Ancylostoma larvae[85]	−	−	−	−	−	−	−	+	−	−	−	−	−	−	−	−
Dictyocaulus larvae[107]	−	⁓	−	−	−	−	−	+	−	−	−	−	−	−	−	−
Nippostrongylus larvae[83,147]	−	−	−	−	−	−	−		−	−	−	−	−	−	+	+
Strongyloides larvae[84]	−	−	−	−	−	−	−		−	−	−	−	−	−	+	−
Animal parasitic fed on tissues																
Ancylostoma[148,149]	+	−	−	−	−	−	−		−	−	−	+	−	−	−	−
Graphidium[44]	−	+	−	−	−	−	−		−	o	−	+	−	−	−	o
Haemonchus[135]	+	−	−	−	−	−	−		−	−	−	+	−	−	−	−
Nippostrongylus[83,98]	−	−	−	+	−	+	+		−	−	−	−	−	−	+	−
Strongylus[117]	−	+	−	o	−	−	−		−	−	−	+	−	−	+	+
Trichuris[99]	−	+	−	+	+	+	+	+	−	−	−	+	−	−	+	o
Animal parasitic fed on gut contents																
Ascaris[22,82,117]	−	+	+	o	o	−	−		−	−	−	+	+	−	+	+
Leidynema[80]	−	+	+	o	o	−	−		−	−	−	+	+	−	?	+

+ = enzyme present
o = enzyme absent
− = not investigated
? = uncertain result

gut contents, although lactase and invertase are not present.[22, 80] Several plant parasitic nematodes possess invertase, but the only animal parasitic nematode found to possess invertase is *Trichuris*.[99] Trehalase is present in homogenates of this species but is probably associated with the nematodes' own carbohydrate metabolism.

The glycosidases of *Trichuris* and of *Nippostrongylus* have been studied in great detail.[98, 99] They show many resemblances to the corresponding enzymes from mammalian sources and are closer to them than to the glycosidases of other invertebrates. The β-acetylaminodeoxyglucosidase, β-glucuronidase, β-galactosidase and α-mannosidase of *Nippostrongylus* are much more active than the corresponding enzymes in mammals. *Trichuris*, however, has very much less β-glucoronidase and β-galactosidase activity than have mammalian tissues. Both of these nematodes feed upon the mucosa of the host, so the presence of high glycosidase activity is significant in view of the suggestion that muco-substances may be natural substrates of many glycosidases.[98, 99]

(b) *Digestion of proteins*

Most of the food of predatory, stylet-bearing nematodes is probably digested extracorporeally, so that little further digestion will take place in the intestine. Carnivorous forms (such as *Mononchus*), which swallow prey whole or in pieces, obviously possess digestive enzymes (including proteases) in the intestine, as digestion of prey can be observed taking place in the living nematode. There is, however, no detailed information on these enzymes in any predatory or microbivorous nematode. Some phytophagous nematodes have protease enzymes although these are not present in *Ditylenchus triformis* (table I).[77, 95]

The skin-penetrating larvae of animal parasitic nematodes possess a collagenase-like enzyme but this is believed to be secreted by the pharyngeal glands (table I).[84]

Nematodes parasitic in animals have proteolytic enzymes in the intestine (table I); the nematodes which feed on the tissues of the host have higher enzyme activity than those which feed on partially digested food and bacteria in the lumen of the intestine.[117] These enzymes fall within the *p*H range of mammalian cathepsins, but activators of mammalian cathepsins give negative or doubtful results with the proteases from nematodes. Several animal parasitic nematodes break down haemoglobin to haematin and protein and must, therefore, possess active proteases in the intestine.[140]

Exopeptidase enzymes are present in animal parasitic nematodes[22, 80, 129] (table I). *Ascaris* contains leucine aminopeptidase, a tripeptidase, alanylglycine peptidase and glycylglycine peptidase.[22] No studies on these enzymes have been done on free-living or plant parasitic nematodes.

(c) Digestion of fats

Fats are digested by lipases, which hydrolyse the esters of higher fatty acids, and by esterases, which hydrolyse shorter chain fatty acids.

Esterases appear to be present in the alimentary system of most nematodes as they are found in predators, plant feeders and animal parasitic forms (table I). The esterases in the intestine of *Ascaris* are similar to mammalian A- and B-types[82] and those of *Trichuris* are similar to mammalian B-type esterase.[99] The esterase in the intestine of the predatory nematode *Actinolaimus* is much weaker than the esterase in the phytophagous nematode *Dorylaimus*.[83] This is probably related to the method of feeding of these two nematodes, the predator probably carrying out much more extracorporeal digestion than the phytophagous nematode.

Lipases are not universally found in the intestine of animal parasitic nematodes (table I). The lipase of *Ascaris* is activated by the bile salt sodium glycocholate and inhibited by sodium taurocholate, but that of *Strongylus* is inhibited by both substances.[117] The lipase of the latter species, which feeds on the tissues of the host, is twelve times more active than the lipase of *Ascaris*, which feeds on the gut contents of the host.[117]

No work has been done on the lipases of free-living or plant parasitic nematodes.

In *Ascaris* most hydrolytic enzyme activity takes place in the anterior intestine and the mid-intestine, and the relative activity of the enzymes is:—peptidases$>$proteases$>$maltase$>$lipase$>$amylase.[22]

It seems probable that nematodes produce the normal hydrolytic enzymes, to a greater or lesser extent, in either the pharyngeal glands or the cells of the intestine or in both these places. Much more work on the nature and the distribution of enzymes is needed, especially on the free-living nematodes, before any safe generalizations can be made. From the small amount of evidence available the digestive enzymes, and their relative activity, can be related to the method of feeding and the nature of the food of the nematode. Table I illustrates the relationship between the nature of the food and the digestive enzymes present in the alimentary tract.

Absorption

Very little is known about the absorption of food materials by nematodes. Most evidence indicates that absorption takes place through the intestine and not through the cuticle. Radio-autographs of *Ascaris*, after immersion in sodium phosphate labelled with ^{32}P, show little absorption by the cuticle but a marked uptake of the phosphate by the intestine.[122] Feeding experiments have shown that *Ascaris* can absorb several simple sugars to produce glycogen in the tissues and can absorb several amino acids to produce body fluid protein.[48, 129]

Feeding experiments with *Ascaris*, using horse haemoglobin and protoporphyrin compounds, showed that the haemoglobin content of the body fluid of the nematode increased in ' fed ' worms. The haemoglobin in the body cavity of these ' fed ' worms was not horse haemoglobin but had all the properties of *Ascaris* haemoglobin. The horse haemoglobin had thus been broken down, absorbed and then reformed as *Ascaris* haemoglobin.[140] Acid and alkaline phosphatases have been demonstrated, histochemically, in the intestine of *Ascaris* and of *Parascaris*[161] and these enzymes are believed to take part in the active absorption of sugars against a concentration gradient. Acid phosphatase is concentrated in the microvilli of the intestine in *Ascaris* and decreases in activity when the nematode is starved.[161]

Certain dyes have been ingested from the medium by some nematodes and passed to the excretory canals from the intestine, showing that absorption has taken place in the intestine.[4] The anterior part of the intestine appears to be primarily secretory and the mid-intestine and the posterior intestine the main absorption areas in several nematodes. In some nematodes, however, digestion is carried out extracorporeally and the intestine is thus wholly absorptive in function. The microvilli greatly increase the absorptive surface of the intestinal cell (in *Ascaris* by a factor of 75 to 90).[72]

Most evidence points to extracellular digestion and absorption of simple food materials in the intestine of nematodes. Evidence that phagocytosis and intracellular digestion take place in the intestine is lacking. *Graphidium strigosum* ingested powdered charcoal, and carbon particles were later found in the cells of the mid- and posterior intestine making them appear grey but very few particles were present in the cells of the anterior intestine.[44] This suggests that some cells may ingest particles from the lumen of the intestine.

The *p*H of the Intestine

The intestinal contents of nematodes have rarely been studied for acid or alkaline reactions. The *p*H of the cells of the intestine of *Diplogaster* and *Pelodera* varies, in different individuals, from about *p*H 4 to 6·8. The mean *p*H in *Pelodera* declines from 6, after being cultured for 6 and 13 days, to about 5·2 in a culture 30 days old. In cultures of *Diplogaster* the mean *p*H declines from 6·2 on the sixth day to 5·5 on the thirteenth day. There is also variation in *p*H along the length of the intestine in *Pelodera*. The posterior region of the intestine usually has a slightly acid *p*H (6·2–6·8). Food in the pharynx shows a *p*H of 6 to 7.[39] The *p*H of the alimentary canal of *Graphidium* is 7 at the anterior and posterior ends and 4·6 in the middle.[44] The contents of the intestinal caecum of *Leidynema*, when released into an indicator solution, give a *p*H value of 5.[83] It is possible that there is an increase in alkalinity along the length of the intestine of *Leidynema* with amylase (*p*H 4·5) being active in the caecum and anterior intestine, and protease and lipase (*p*H 6 and 7) being active in the middle and posterior intestine.[80] The intestinal contents of *Ascaris* must be alkaline or near to neutral as optimum activity of the digestive enzymes takes place between *p*H 6 and 9·4.[117] As the digestion of carbohydrates and fats proceeds, however, the intestinal contents may become more acid thus allowing the protease (*p*H 6) to operate. The intestinal contents of *Strongylus* are about *p*H 8.[117]

3: Metabolism and Oxygen Transport

Distribution of Carbohydrates

In most nematodes glycogen provides the major reserve store of energy. This applies particularly to nematodes which are parasitic in animals and which live in environments with low oxygen tensions. The glycogen is stored mainly in the hypodermis, the non-contractile part of the muscles, the intestine and in the epithelial cells of the reproductive organs.

The amount of glycogen in the tissues of nematodes varies from species to species and also within the species, but it is usually present in considerable amounts. In *Ascaris* the amount of glycogen present is about 20 per cent. of the dry weight of the animal, and about 70 per cent. of the dry weight of the muscle is composed of this carbohydrate.[48] Many free-living stages of animal parasitic nematodes have very little stored glycogen, the energy reserve being mostly fat. Free-living and plant parasitic nematodes have been little studied, but they apparently store less glycogen than the animal parasitic species. This is probably related to their predominantly aerobic existence which means that they are able to use stored fats.

The only glycogen which has been studied in detail in nematodes is that of *Ascaris*, in which the molecular weight of the glycogen is much greater than that of mammalian glycogen.[18]

Glycogen disappears rapidly from the tissues of nematodes when they are not fed. In *Ascaridia galli* there is a rapid decline in the amount of glycogen present in the tissues of the nematode when the host (the fowl) is not fed, and after three days the

nematodes are expelled from the host. *Ascaris* consumes 1·4 gm.
of glycogen per 100 gm. body weight in 24 hours under anaerobic
conditions and 0·85 gm. under aerobic conditions.[12]

The anthelmintic dithiazine interferes with the uptake and
transport of glucose in *Trichuris vulpis* at concentrations which do
not affect the mobility of this animal parasitic nematode. As a
result of this inability to use exogenous glucose, the endogenous
stores of carbohydrate are utilized and there is a decrease in the
generation of energy-rich phosphate bonds which eventually
causes the death of the nematode.[20]

Sugars are also present in the tissues and in the body fluid of
nematodes. In *Ascaris* the intestine and the body wall contain
more glucose than trehalose, the disaccharide is, however, present
in greater amounts in the muscles, reproductive organs and the
pseudocoelomic fluid. Trehalose is apparently the most abundant
low molecular weight carbohydrate in nematodes (table II).[48]
Glucose and trehalose, which comprise 5–7 per cent. of the dry
weight, form reserves which are used extensively when *Nippostrongylus brasiliensis* is starved in saline.[113]

TABLE II

*The glucose and trehalose content of several animal parasitic
nematodes*[48]

	% of tissue solids	
Nematode	Trehalose	Glucose
Porracaecum decipens (larvae)	2·18	0·16
Trichinella spiralis (larvae)	1·76	0·04
Uncinaria stenocephala	0·91	0·77
Trichuris ovis	0·48	0·09
Ascaridia galli	0·38	0·78
Heterakis gallinae	0·10	0·43
Litomosoides carinii	0·06	0·01

Ascarylose (3,6-di-deoxyaldohexose) is a hexose found only
in nematodes. In *Parascaris* it occurs in the oöcytes where it
contributes to the formation of the inner membrane of the egg
shell.[48] In nematodes, chitin is found only in the egg shell and is
synthesized by the egg and not by the reproductive system of the
female.

D—P.N.

TABLE III

Enzymes of the Embden-Meyerhof pathway which have been demonstrated in nematodes

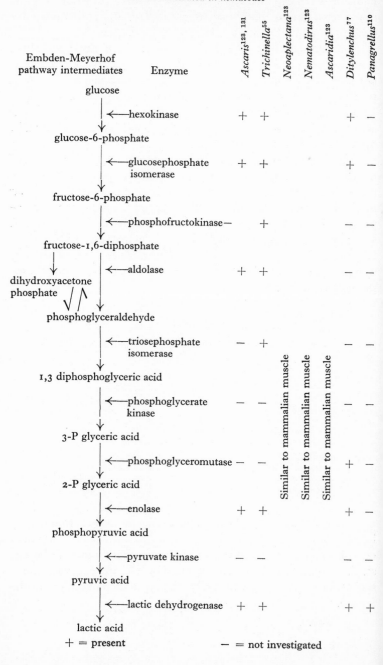

Embden-Meyerhof pathway intermediates	Enzyme	Ascaris[128,131]	Trichinella[55]	Neoplectana[123]	Nematodirus[123]	Ascaridia[123]	Ditylenchus[77]	Panagrellus[110]
glucose	hexokinase	+	+				+	−
glucose-6-phosphate	glucosephosphate isomerase	+	+				+	−
fructose-6-phosphate	phosphofructokinase		+				−	−
fructose-1,6-diphosphate / dihydroxyacetone phosphate	aldolase	+	+					
phosphoglyceraldehyde	triosephosphate isomerase	−	+	Similar to mammalian muscle	Similar to mammalian muscle	Similar to mammalian muscle	−	−
1,3 diphosphoglyceric acid	phosphoglycerate kinase	−	−				−	−
3-P glyceric acid	phosphoglyceromutase	−	−				+	−
2-P glyceric acid	enolase	+	+				+	−
phosphopyruvic acid	pyruvate kinase	−	−				−	−
pyruvic acid	lactic dehydrogenase	+	+				+	+
lactic acid								

+ = present   − = not investigated

Carbohydrate Metabolism

In most animal cells carbohydrate is broken down by a combination of the Embden-Meyerhof pathway (glycolysis) and the aerobic tricarboxylic acid (TCA or Krebs') cycle. The Embden-Meyerhof pathway converts glucose or glycogen to pyruvate (table III). The resultant pyruvic acid may then be reduced to lactic acid or may enter the TCA cycle (fig. 13).

The TCA cycle oxidizes not only carbohydrate but also most of the fat and some of the protein intermediates of the cell. The sequence of reactions produces $NADH_2$, $FADH_2$ and GTP. GTP can react with adenosine diphosphate (ADP) to form GDP and adenosine triphosphate (ATP), which contains 'high energy bonds'. The subsequent oxidation of $NADH_2$ also leads to the production of energy.

Aerobic metabolism of glucose can yield thirty-eight 'high energy' phosphates, whereas anaerobic glycolysis yields only two.

Embden-Meyerhof pathway

Nematodes parasitic in animals obtain some of their energy from carbohydrates by anaerobic processes and the evidence suggests that the Embden-Meyerhof pathway functions in many of these nematodes. Table III shows the nematodes in which enzymes involved in the Embden-Meyerhof pathway have been demonstrated. Little is known about carbohydrate metabolism in free-living nematodes but enzymes of the Embden-Meyerhof pathway are present in a plant parasitic nematode (*Ditylenchus*).

Oxygen debt

In vertebrates, when oxygen is not available, energy is released by the formation of lactic acid from glucose. This lactic acid is subsequently metabolized when oxygen becomes available, and there is higher than normal oxygen consumption while the lactic acid is metabolized. This is called repayment of an oxygen debt.

Many nematodes are subjected to prolonged periods of anaerobiasis, or normally live in low oxygen tensions, and obtain much of their energy by the anaerobic breakdown of carbohydrate. They excrete some, or all, of the organic acids which are formed,

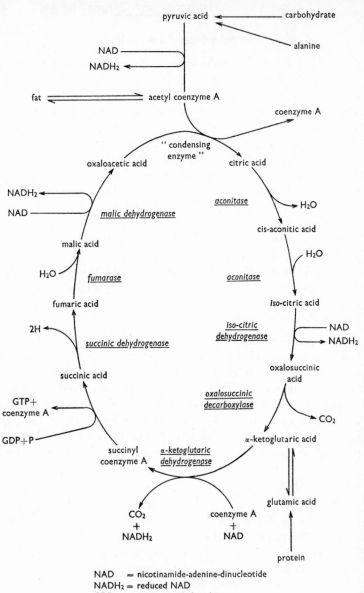

FIG. 13. Tricarboxylic acid cycle.

however, and so have a reduced oxygen debt to repay. Thus, *Litomosoides* shows no postanaerobic increase in oxygen consumption, as the carbohydrate consumed under anaerobic conditions is equal to the amount of acid excreted,[16] while *Ascaris* and larvae of *Eustrongyloides* show only small postanaerobic increases in oxygen consumption.[12, 48]

End-products of carbohydrate metabolism

Lactic acid is not necessarily the end-product of carbohydrate catabolism in nematodes, in fact it is only excreted in appreciable quantities by three of those species which have been investigated (*Litomosoides*, *Dracunculus* and *Dirofilaria*).[14, 16] It is produced in small or trace amounts by a few other nematodes (table IV).

Other organic acids and products from carbohydrate catabolism which are excreted by nematodes are shown in table IV. Pyruvate appears to be essential for the formation of these organic acids.

TABLE IV

The excretory products formed from the breakdown of carbohydrate by some nematodes

Excretory product	Ascaris[42]	Trichinella larva[15]	Heterakis[48]	Litomosoides[16]	Dracunculus[21]	Dirofilaria[14]	Caenorhabditis[127]	Ancylostoma[32]	Trichuris[20]
lactic acid	T	T	T	+	+	+		−	+
propionic acid	+	+	+	−	−	−		+	+
acetic acid	+	+	+	+	−	−		+	−
pyruvic acid	−	−	+	−	−	−		−	−
succinic acid	−	−	+	−	−	−		−	−
α-methylbutyric acid	+	−	−	−	−	−	Only trace amounts	+	−
n-valeric acid	+	+	−	−	−	−		−	+
iso-caproic acid	+	−	−	−	−	−		−	−
n-caproic acid	+	+	−	−	−	−		−	−
acetylmethyl carbinol	+	−	−	+	−	−		−	−
iso-butyric acid	−	−	−	−	−	−		+	−
n-butyric acid	+	+	−	−	−	−		−	T
cis-d-methylcrotonic (tiglic) acid	+	−	−	−	−	−		−	−
C6 acids (unidentified)	−	+	−	−	−	−	T	−	−

T = trace amount only
+ = present
− = not investigated or absent

Table IV shows that some nematodes excrete a more complex mixture of organic acids than others. The composition of the acids excreted by some nematodes (*Trichuris, Trichinella, Heterakis* and *Dracunculus*) is not affected by the presence or absence of oxygen;[15, 20, 48] *Litomosoides*, on the other hand, converts 80 per cent. of the carbohydrate metabolized to lactic acid, together with small amounts of acetic acid, under anaerobic conditions. Under aerobic conditions, however, only 30–45 per cent. of the glucose is converted to lactic acid—larger amounts of acetate and some acetylmethyl-carbinol being produced.[16]

When *Ascaris* is given glucose labelled with radioactive carbon, the ^{14}C is incorporated into glycogen and the excreted organic acids. There is also a relatively high incorporation of ^{14}C from glucose into protein.[46]

Caenorhabditis is the only free-living nematode which has been studied and it produces only trace amounts of volatile acids.[127]

Carbon dioxide is produced by all nematodes and is usually excreted through the general body surface of the nematode. In some nematodes, however, it is used in metabolism. In *Heterakis*, in *Ascaris* and possibly in *Trichuris* metabolically produced carbon dioxide is used by the nematode to produce succinate from pyruvate and eventually the carbon from this carbon dioxide appears in the organic acids which are excreted by the nematode.[20, 48, 130]

Hexose mono-phosphate shunt

The hexose mono-phosphate shunt has been looked for in only a few nematodes. Glucose-6-phosphate dehydrogenase and gluconate-6-phosphate dehydrogenase are present in several animal parasitic nematodes[37] and in the plant parasitic nematode *Ditylenchus*.[77] These two enzymes are present in relatively high amounts but the evidence indicates that the hexose mono-phosphate shunt is probably relatively unimportant as a source of energy in these nematodes. In *Ascaris* it may take part in the synthesis of nucleic acid pentose and in the formation of reduced NADP, which is required for lipid synthesis.[46]

Tricarboxylic acid cycle

Some nematodes possess all the components of the TCA cycle but others apparently possess only a part of the cycle (table V).

TABLE V

Enzymes of the tricarboxylic acid cycle which have been demonstrated in nematodes

TCA cycle intermediates	Enzyme	Turbatrix[43]	Panagrellus[110]	Ditylenchus[77]	Neoaplectana[89]	Nippostrongylus larvae[134]	Necator larvae[51]	Strongyloides larvae[30]	Ascaris[109]	Trichinella[54]	Ascaridia[89]	Nematodirus[89]
oxaloacetic acid												
←—'condensing enzyme' citrate synthase		+	−	−	+	−	−	−	o	+	+	+
citric acid												
←—aconitase		+	+	−	−	−	−	+	o	+	+	+
isocitric acid												
←—isocitric dehydrogenase		o	−	+	−	−	−	+	o	+	−	−
oxalosuccinic acid												
←—oxalosuccinic decarboxylase		o	−	+	−	−	−	+	o	+	−	−
ketoglutaric acid												
←—a-ketoglutaric dehydrogenase		−	−	o	+	+	−	?	+	+	+	+
succinic acid												
←—succinic dehydrogenase		?	+	?	+	+	+	+	+	+	+	+
fumaric acid												
←—fumarase		?	−	+	+	−	−	−	+	+	+	+
malic acid												
←—malic dehydrogenase		+	−	+	+	−	−	+	+	+	+	+
oxaloacetic acid												

? = presence doubtful   + = present
− = not investigated   o = absent

Panagrellus silusiae, a free-living nematode, appears to have a complete TCA cycle but *Turbatrix aceti*, which is also free-living, does not possess the TCA cycle although it has a predominantly aerobic metabolism.[43, 110]

The only two plant parasitic nematodes which have been studied (two species of *Ditylenchus*) contain several enzymes of the TCA cycle indicating that at least a partial TCA cycle is operating.[77]

Most of the work on the TCA cycle in nematodes has been done on animal parasitic nematodes and, while some appear to have the ability to metabolize all, or most, of the TCA cycle substrates, others do not have a functional TCA cycle (table V). The occurrence of most of the cycle has been demonstrated in four species (*Trichinella, Nematodirus, Neoaplectana* and *Ascaridia*) and these presumably metabolize pyruvate by this mechanism.[54, 89]

Electron transport system

Cytochromes and cytochrome oxidase are found in some nematodes, chiefly free-living and plant parasitic species, but the terminal oxidation of animal parasitic nematodes seems to have undergone marked changes. *Litomosoides*, which lives in the pleural cavities of vertebrates, is dependent upon oxygen for its metabolism but it has no cytochrome *c* or cytochrome oxidase.[19] *Ascaris* possesses neither cytochrome *c* nor cytochrome oxidase,[19] instead the electron transport system of *Ascaris* muscle reacts with oxygen by means of a flavin-containing terminal oxidase.[18] There may be a cyanide insensitive cytochrome *b* system in this nematode[73] but this is questionable. Cytochrome *c* and cytochrome oxidase are present in *Trichuris vulpis* but, as the metabolism of the worm is essentially anaerobic, the cytochrome system may play no physiological rôle.[20] Cytochrome oxidase and cytochrome *c* are present in both the larvae and adults of *Trichinella* which also possess the complete TCA cycle.[54] Carbohydrate metabolism in this nematode appears to be similar to that in most other animals. Cytochrome *c* is present in the larvae of *Nippostrongylus*[83] and cytochrome oxidase is also plentiful in the adult so this may be the main electron transport system. Flavin-containing systems are not important in this nematode.[113]

The cytochrome *c*-cytochrome oxidase electron transport

system and the enzymes of the TCA cycle are associated with the mitochondrion. Mitochondria, of normal appearance, are present in the tissues of most nematodes which have been studied with the electron microscope (*Nippostrongylus brasiliensis* larvae and adult;[83] *Capillaria hepatica*).[159, 160] *Ascaris* and *Parascaris*, however, have mitochondria with very few short cristae.[62, 75] The small number of cristae in *Ascaris* mitochondria may be related to the lack of a cytochrome *c*-cytochrome oxidase electron transport system and to the lack of a complete TCA cycle in the nematode.[75]

Metabolism in Ascaris

The intermediary metabolism of *Ascaris* is rather unusual and will therefore be dealt with separately.

Ascaris possesses all the enzymes of the Embden-Meyerhof pathway of glycolysis[131] and glucose is metabolized to pyruvate by this route. Lactic dehydrogenase activity is low which explains why only trace amounts of lactic acid are excreted by this nematode. The TCA cycle seems to play no part in *Ascaris* metabolism, for most of the enzymes associated with the TCA cycle are not present. Other pathways must, therefore, be present to reoxidize the reduced NAD (nicotinamide-adenine-dinucleotide) which is formed during glycolysis. This can be brought about by the production of fumarate from pyruvate (fig. 14). *Ascaris* is able to produce fumarate, anaerobically, by carbon dioxide fixation into pyruvate.[130] The fumarate is then reduced to succinate with the reoxidation of more reduced NAD.[74] The succinic dehydrogenase system in *Ascaris* catalyzes the reduction of fumarate by $NADH_2$ (reduced NAD) to produce succinate much more rapidly than it catalyzes the dehydrogenation of succinate. It seems then that this reaction plays an important rôle in the electron transport system in *Ascaris* muscle.[131] The reduction of fumarate by $NADH_2$ suggests that electrons from $NADH_2$ may be directed either to oxygen or, under anaerobic conditions, to fumarate. The succinoxidase system of *Ascaris* can therefore serve as an electron acceptor from succinate or as an electron donor to fumarate. Under anaerobic conditions the formation of succinate in *Ascaris* would reoxidize the reduced NAD formed during glycolysis and thus supply energy for muscular

contraction.[18] When *Ascaris* is paralyzed by the anthelmintic drug,
' Piperazine ', there is a marked diminution in the formation of
succinate.

The yield of ATP for the oxidative steps of the TCA cycle is
evaluated by measuring the ratio of inorganic phosphorus incorporated to oxygen used (P/O ratio). The P/O ratio, when
α-ketoglutarate is the substrate, is 4; it is 2 when succinate, 3 when
malate and 3 when isocitrate are the substrates. Aerobic oxidative

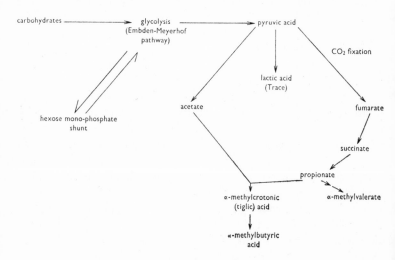

FIG. 14. Possible metabolic pathway in *Ascaris*.

phosphorylations occur in washed homogenates of *Ascaris* muscle
without any added substrate and give P/O ratios of 1·3 to 3·5
Washed particles from *Ascaris* muscle catalyze aerobic phosphorylations when pyruvate, catalase and dialyzed pseudocoelomic
fluid are present, but the P/O ratios are very low. When succinate
is the added substrate, or if undialyzed pseudocoelomic fluid
(which normally contains a high concentration of succinate) is
used without added substrate, oxidative phosphorylation is reduced
even further. Thus, there is apparently no direct coupling
between oxidative phosphorylation and the oxidation of succinate
in *Ascaris*.[122]

Ascaris eggs, first- and second-stage larvae, unlike the adult worm, contain the cytochrome c-cytochrome oxidase system of electron transport.[103]

The intermediary metabolism of *Trichuris vulpis* may be similar to that of *Ascaris*. The nematode is not dependent on aerobic metabolism and, in an atmosphere which contains 2–5 per cent. carbon dioxide in nitrogen it survives for considerably longer periods and metabolizes carbohydrate at a higher rate than in nitrogen alone, in air, or in mixtures of carbon dioxide and oxygen showing that energy is provided by anaerobic rather than aerobic metabolism. The marked effect of carbon dioxide in increasing the period of survival and in enhancing the rate of glucose uptake may be due to a mechanism for carbon dioxide fixation similar to that which occurs in *Ascaris*.[20]

Production of organic acids in Ascaris

Ascaris ferments glucose to carbon dioxide and a mixture of excretory products. The major excretory product of carbohydrate catabolism is α-methylbutyric acid which accounts for 20 per cent. of the volatile acids produced. Valeric acid probably accounts for 10–15 per cent. Some of the succinate produced by *Ascaris* is excreted but some is decarboxylated to form propionate which then forms α-methylcrotonic and α-methylbutyric acids by condensation with acetate and these are then excreted (fig. 14). The acetate required for this reaction comes from the oxidative decarboxylation of pyruvate.[131] Propionate is also the precursor of α-methylvalerate which is another excretory product of *Ascaris* (fig. 14).[48, 122, 131]

Distribution of Fats

Glycogen is the chief food reserve in most nematodes but considerable amounts of fat are also stored, especially by free-living species and by the free-living stages of animal parasitic nematodes.

Most of the fat is located in the hypodermis, the noncontractile part of the muscle cells, in the intestine and in the ovaries. In many free-living stages of animal parasitic nematodes fat is stored in the lumen of the intestine, especially in those stages which do not feed.

The amount of fat present in nematodes varies widely and is

usually much less in animal parasitic nematodes than in free-living or plant parasitic species. This is probably related to the amount of oxygen available in the environment. The amount of lipid present in two animal parasitic nematodes (*Dioctophyme renale* and *Ascaris*) is 1 per cent. and 8 per cent. respectively of the dry weight whereas in a free-living nematode (*Turbatrix aceti*) it is 41 per cent. and in a plant parasitic species (*Ditylenchus*) 33 per cent.[48]

Detailed studies on the nature of the lipids of nematodes have usually been confined to *Ascaris* and *Parascaris* making it impossible to generalize on the nature of the lipids found in nematodes. The nature of the lipids and their distribution in the tissues of female *Ascaris* are given in table VI.

TABLE VI

The nature of the lipids and their distribution in female Ascaris[48]

Tissue	(% wet weight) Total lipid	(% of total lipids) Tri-glycerides	Phospho-lipids	Un-saponifiables
Pseudocoelomic fluid	0·32		62	
Cuticle	0·61	43	41	16
Muscle	0·75	40	49	11
Reproductive system	6·0	70	8·3	22
Fertilized eggs	17·5	69	9·2	22

The phospholipids and the unsaponifiables are probably structural components but the triglycerides serve as an energy reserve as well as being structural lipids.[48] The unsaponifiables of *Ascaris* contain cholesterol, a second saturated sterol and ascaryl alcohol.[47] Ascaryl alcohol from the egg membranes of *Parascaris* is a mixture of three glycosides called ascaroside A, B and C. They are esterified with acetic and propionic acid in the unfertilized egg. Ascarosides occur in the muscle and cuticle of *Ascaris*[47] and in *Eustrongyloides* but they are most abundant in the oöcytes and in the fertilized eggs. For further information on the nature of nematode lipids see Fairbairn.[47, 48]

Fat Metabolism

The fat metabolism of nematodes has been little studied and most of the detailed work has been done on animal parasitic species.

The free-living, non-feeding, infective larvae of several animal parasitic nematodes (*Nippostrongylus, Haemonchus, Ancylostoma, Necator*) use their stored fat as an energy reserve. The

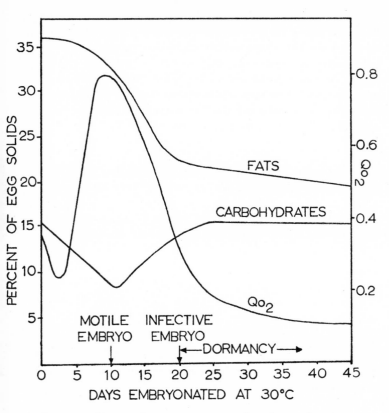

FIG. 15.   Changes in oxygen consumption, fats and carbohydrates during embryonation of *Ascaris* eggs (after Fairbairn[48]).

infectivity of these larvae is related to the amount of the stored fat.   Larvae which have used up their reserves of fat are not as infective, nor as motile, as larvae containing large amounts of fat.[122]

In nematodes, as for other animals, the breakdown of endogenous fats requires oxygen.   Larvae of *Trichinella spiralis*, when kept in non-nutrient media under anaerobic conditions, do not use any of their stored fat, but when oxygen is admitted there is a significant decrease in the amount of stored fat and this is independent of the carbohydrate consumed.[12]   It is thought that the nematodes use these fats as a source of energy for movement.[12] An insect parasitic nematode (*Thelastoma*) makes use of its stored fat when not fed for periods of up to 12 days but adult *Ascaris*, which have not been fed, are apparently unable to utilize their stored fat to any great extent.[12]   Nematodes (including *Ascaris*) must have an active fat metabolism, however, as fats are deposited in the oöcytes and act as a food reserve for the egg, in which there is a definite fat catabolism.   The unembryonated eggs of *Ascaris* contain large amounts of fat which disappear slowly at first and then more rapidly as the egg develops in the presence of oxygen (fig. 15).   The infective larva inside the egg uses fat much more slowly once development is completed.   When fat is being used most rapidly by the developing egg the oxygen consumption actually begins to decrease, which shows that only part of the fat is being completely oxidized and the rest is being converted to carbohydrate.   As a result, the total amount of carbohydrate after 25 days at 30° C. is about the same as in the unembryonated egg but the amount of fat is greatly reduced.[48]

No work has been carried out on the free-living or plant parasitic nematodes but it is probable that most species live aerobically and have a significant fat metabolism.

Distribution of Proteins

Proteins are the chief structural components of the cuticle, muscles and other tissues of nematodes, and the intestine of several nematodes contains colourless globules of protein material,[26] but there is little evidence that protein is stored as an energy reserve. Protein granules in the intestine of larval *Agamermis* and *Meloidogyne* disappear as the larvae develop and may be used to form structural proteins as the nematode increases in size.[26, 27]

Haemoglobin is present in several animal parasitic nematodes and, although in the majority of cases it is probably associated with the uptake of oxygen by the nematode, there is strong evidence that this is not always so. The haemoglobin found in the pseudo-coelomic fluid of *Ascaris* is not associated with the storage or transport of oxygen but acts as a source of haem which is incorporated into the eggs of the nematode.[140]

Protein Metabolism

The metabolism of nitrogen in nematodes has been little investigated. In adult nematodes most of the nitrogen anabolism is concerned with the formation of the eggs. This is not surprising when it is realized that a single female *Ascaris* lays 200,000 eggs a day and many other animal parasitic species are equally prolific. Most of the proteins deposited in the oöcyte become structural components of the larva and the egg shell and are not used as a source of energy. It is possible that some amino acids required by the gonads are derived from carbohydrates as well as from amino acids in the food of the nematode, because *Ascaris* ovaries can form the amino acid alanine from pyruvate and ammonia.[106]

The free-living nematode, *Caenorhabditis briggsae*, is able to synthesize certain ' essential' amino acids (valine, lysine, threonine, leucine and isoleucine), an ability formerly attributed only to plants and micro-organisms.[128]

There is very little information on the uptake of proteins by nematodes. Amino acids added to a medium containing starved *Ascaris* bring about an increase in the amount of protein in the pseudocoelomic fluid.[129] Also horse haemoglobin, when fed to *Ascaris*, results in an increase in the pseudocoelomic fluid haemoglobin of the nematode. The pseudocoelomic fluid contains no trace of horse haemoglobin after this feeding showing that the haemoglobin has been broken down and resynthesized by the worm.[140]

Excretion of nitrogen

The nitrogenous excretory products of nematodes are shown in table VII. Ammonia is the chief excretory product of protein metabolism in nematodes and it is excreted through the body wall and through the intestine. The ammonia is probably produced by the deamination of amino acids as in vertebrates; in fact this process has been demonstrated in three animal parasitic nematodes (*Nematodirus, Ascaridia* and *Ascaris*).[24, 120, 129] In *Ascaris* amino acid oxidase activity is greatest in the intestine and this is

TABLE VII

The nitrogenous excretory products excreted by some fasting nematodes in vitro[155]

Nematode	Total nitrogen excreted mg./gm. wet wt. 24 hr.	Type of nitrogen, % of total					
		ammonia	urea	peptide	amino acid	amine	uric acid
Caenorhabditis[127] (Free-living)	—	+	o	—	+	T	o
Ascaris							
aerobic	0·4	69	7	21	+	—	o
anaerobic	0·4	71	6	18	+	—	o
In U-tube	0·4	27	51	19	—	—	o
larvae						+	—
Nematodirus							
aerobic	1·36	42	14	35	+	—	3
anaerobic	1·72	29	4	35	+	—	4
Ascaridia							
aerobic	0·35	56	12	15	+	—	o
anaerobic	0·37	59	15	15	+	—	o
Trichinella (larva)							
aerobic	2·8	33	o	21	28	7	o
Nippostrongylus (3rd larva)							
aerobic	—	+	o	—	—	+	—

+ = present
o = absent
— = no information
T = trace amounts

presumably the main site of ammonia formation in this nematode. Nematodes are aquatic organisms and so the ammonia excreted into the environment will be rapidly diluted by the surrounding water.

Urea, which is excreted by some nematodes (table VII), is formed from ammonia by the ornithine cycle.[120] This cycle can be described rather simply as follows:—Ornithine, together with carbon dioxide and ammonia produce citrulline. More ammonia combines with the citrulline to become arginine which combines with water to produce urea and ornithine. This last reaction is catalyzed by arginase and has been demonstrated in several nematodes (*Ascaris, Ascaridia* and *Nematodirus*). This enzyme is especially concentrated in the intestine of *Ascaris*, indicating that this is the main site of urea production. The breakdown of purines also results in the formation of urea and ammonia in *Ascaridia* and *Ascaris*.[120] This probably takes place as follows:—

$$\text{purine} \xrightarrow{\text{uricase}} \text{uric acid} \xrightarrow{\text{allantoinase}} \text{allantoin} \xrightarrow{} \text{allantoic acid}$$

$$\xrightarrow{\text{allantoicase}} \text{glyoxylic acid} + \text{urea} \xrightarrow{\text{urease}} \text{ammonia} + \text{carbon dioxide}.$$

Several nematodes possess a urease which converts urea to ammonia; this is rather rare in animals. The substrate for the enzyme may be provided by ingested urea or urea produced by the nematode.

Some free-living and parasitic nematodes excrete large amounts of amino acids. Many other invertebrates excrete amino acids but it is not known whether this is just a simple leakage from the cells of the animal, because of their incubation in abnormal media, or whether it is a natural phenomenon.[108] Peptides are also excreted by many nematodes but unfortunately the source of the excreted amino acids and peptides is not known.[155]

Amines are not normally excreted by animals but they are excreted by larvae of several animal parasitic nematodes (table VII).[155] The amines are excreted through the excretory pore of *Nippostrongylus* larvae and are also found in the fluid bathing the larvae in *Ascaris* eggs.[48, 155] This would suggest that these amines are normal excretory products. Decarboxylation of amino acids could produce some of these amines.[122]

The amount of nitrogenous material excreted by an animal indicates the amount of protein in the diet used in the production of energy. *Nematodirus*, which feeds on the intestinal tissues of the host, excretes much more nitrogenous waste than *Ascaridia*, which feeds on the contents of the intestine. This suggests that *Nematodirus* uses protein as a source of energy to a greater extent than *Ascaridia* which relies more on carbohydrate.[120]

Uptake of Oxygen

There is no circulatory system in nematodes, so oxygen must travel from the environment across the body wall to the internal organs by diffusion through the pseudocoelomic fluid. This process may be helped by locomotory movements of the nematode which bring about movements of the fluid, and possibly also by respiratory pigments in some animal parasitic nematodes.

Nematodes which feed on blood (*Ancylostoma*) may extract oxygen from the blood of the host through their intestine.

In larvae and in most small nematodes, diffusion will be sufficient to supply oxygen to the internal tissues because of the greater surface/volume ratio of the animals. *Nippostrongylus* adults, which live in the mucosa of the rat intestine, can obtain sufficient oxygen to carry out aerobic metabolism without an efficient oxygen transporting system.[122] With increase in size, however, diffusion alone may not suffice to supply sufficient oxygen for aerobic metabolism to the central tissues. Thus *Nematodirus filicollis* and *Haemonchus contortus* obtain by diffusion only sufficient oxygen for 50 per cent. or less of their needs. In still larger nematodes, such as *Ascaris*, which is a facultative aerobe, an efficient circulatory system would be required to carry sufficient oxygen to the inner tissues if complete aerobic metabolism were to take place.[122]

Haemoglobin

Those nematodes which live in environments with low oxygen tensions would be expected to have respiratory pigments with low loading tensions, thus enabling them to obtain oxygen necessary for their metabolism. Haemoglobin is the only respiratory pigment which has been found in nematodes. It is present in some animal parasitic nematodes but has not yet been recorded from any free-living or plant parasitic nematode. When haemoglobin is present it is contained in the pseudocoelomic fluid or in the tissues, or in both these places. In every nematode which has been investigated in detail the haemoglobin is nematode haemo-globin and not that of the host. In *Ascaris*, mammalian haemoglobin can be broken down and absorbed, resulting in an increase in the amount of *Ascaris* haemoglobin present in the pseudocoelomic fluid.[140]

Haemoglobin is usually concerned with the supply of oxygen to the tissues. In many animals there are two types of haemo-globin present: one in the blood, which transfers oxygen, and the other (myoglobin) in the tissues, which stores oxygen. The presence of haemoglobin in tissues may, however, facilitate the passage of oxygen through the tissue.[132] In nematodes the properties of haemoglobins have been studied in only a few species. In some of these the haemoglobin contains 1 haem group per protein molecule and is thus similar to vertebrate myoglobin.

Different haemoglobins become saturated with oxygen at different partial pressures of oxygen and this seems to have an adaptive significance. The haemoglobins of some animals, which live in oxygen deficient environments, have lower loading tensions than related animals which live in high oxygen tensions.

Many nematodes which live in the alimentary tract of animals, that is in an environment containing little oxygen, have haemoglobin with a high affinity for oxygen (fig. 16). They can thus extract oxygen from the tissues of the host but the oxygen tension of their own tissues will have to be extremely low before the oxyhaemo-globin gives up the oxygen. The oxyhaemoglobins of such species (*Camallanus*, *Eustrongyloides* larvae, *Heterakis*, *Nippostrongylus*, *Haemonchus* and *Nematodirus*) become deoxygenated under anaerobic conditions and reoxygenation occurs when air is admitted, indicating that these haemoglobins have a respiratory function.[12, 34, 59, 119] This may not be so in all of these nematodes as the haemoglobins of some of them (*Nippostrongylus*, *Haemonchus* and *Nematodirus*) can be poisoned with carbon monoxide without decreasing the oxygen uptake.[118] It is possible, however, that the haemoglobins of these nematodes, which have a low affinity for carbon monoxide, can be effective transporters of oxygen at extremely low tensions of oxygen.[122]

Ascaris possesses two haemoglobins, one in the body wall and the other in the pseudocoelomic fluid. The body wall haemo-globin seems to function as a myoglobin, releasing oxygen under anaerobic conditions.[33]   Apparently the pseudocoelomic fluid haemoglobin does not have a respiratory function, as it does not become deoxygenated under anaerobic conditions and is only

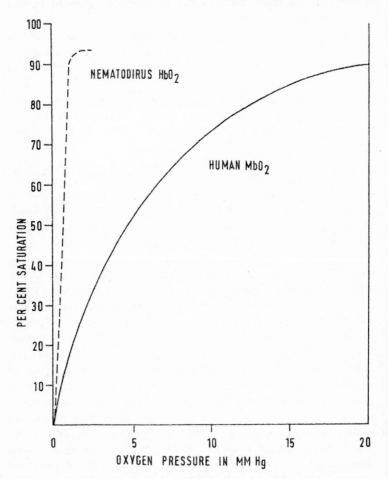

FIG. 16. Comparison of the affinity of *Nematodirus* haemoglobin and human myoglobin for oxygen (partly after Fairbairn[48]).

deoxygenated with difficulty by chemical methods (sodium dithionite). The haemoglobin in the pseudocoelomic fluid of *Strongylus* is similar to that of *Ascaris* in this respect.[34]

The haemoglobin in the body wall of nematodes may not only serve to store oxygen for periods of anoxia but may also function in the transfer of oxygen from the cuticle to the body fluid.[132] Haemoglobin is present in the fluid layer of the cuticle of *Nippostrongylus*;[83] probably it extracts oxygen from the mucosa of the host and passes it to the body wall haemoglobin which in turn passes it to the pseudocoelomic fluid. Thus, the haemoglobins of some nematodes function as oxygen carriers but in others their function is uncertain or not known.

In vertebrate haemoglobins acidification increases the amount of oxygen which dissociates (the Bohr effect). This is physiologically important as an increase in the amount of carbon dioxide in the tissues increases the amount of oxygen which is released by the haemoglobin. Several invertebrate haemoglobins and myoglobin show a small or a negative Bohr effect. In *Ascaris*, acidification brings about a decrease in the dissociation of oxygen from the pseudocoelomic fluid haemoglobin,[33] and maximum dissociation of oxygen from the body wall haemoglobin takes place at pH 7·0.[139] The reverse Bohr effect may be of adaptive value to nematodes if they live under conditions of high carbon dioxide and low oxygen tensions, such as in the vertebrate alimentary canal. The haemoglobin will become loaded with more oxygen in the presence of high carbon dioxide and low oxygen tensions than in low carbon dioxide and low oxygen tensions. If the cellular enzymes of these organisms function at extremely low partial pressures of oxygen (1–2 mm.) this mechanism could be even more important.[111]

Oxygen in the Environment[13]

Many habitats which are completely devoid of oxygen are inhabited by nematodes and these must, therefore, obtain all of their energy requirements from fermentations. The mud at the bottom of lakes and seas, and water-logged soils are often completely devoid of oxygen and yet support significant nematode populations. Mud in the Clyde Sea Area (Scotland) forms an anaerobic habitat and nematodes live in this mud to a depth of 7 cm. An experiment carried out with these marine nematodes showed that they were still alive and active after 35 days in a sealed tube containing mud devoid of oxygen.[92]   The fact that these nematodes were not only alive but active would indicate that they normally have an anaerobic metabolism.   Many nematodes can withstand periods of anaerobiasis but remain quiescent until oxygen is readmitted to the medium.   The number of species of nematodes which are obligate anaerobes is probably very small. Many nematodes are facultative aerobes, however, as they lead an essentially anaerobic existence but can use oxygen if it is available.   Included in this group are some mud inhabiting forms and many animal parasitic nematodes which inhabit the alimentary tract of the host.   These species obtain most of their energy from anaerobic metabolic processes, although they can use oxygen in cellular respiration if it becomes available.   They are often sensitive to oxygen tensions similar to those found in the atmosphere, and one way of ridding patients of *Ascaris* is to pass oxygen into the intestine of the host.

The amount of water in soil directly affects the amount of oxygen available to soil inhabiting nematodes.   In soils saturated with water (fig. 17) the number of pathways between soil particles not blocked with water will be small and the amount of air in the soil will be small.   Also, the depth of water covering nematodes situated in pore spaces in the soil will be much greater in saturated soils than in soils with a low moisture content, therefore oxygen will have much further to diffuse from the air-water interface to the nematode.   This will result in a much slower diffusion of oxygen to the nematode.   The greater the distance oxygen has to diffuse to reach the nematode the more the oxygen will be consumed by other micro-organisms in the soil moisture, and this naturally leads to low oxygen tensions and possibly anaerobic conditions in the water surrounding the nematode.[152]

Soil and plant parasitic nematodes can withstand low oxygen concentrations, but this ability varies from species to species. Decreased oxygen tensions, brought about by the addition of 400 p.p.m. of sodium sulphite to the medium, stopped the movement of *Aphelenchoides* in 4 minutes, *Rhabditis* and *Meloidogyne* in 15 minutes, *Heterodera* in 20 minutes and *Dolichodorus* in 45 minutes.   Recovery of motility after aeration of the solution occurred within 2 to 5 minutes.   Aeration of the soil also has an effect upon hatching of plant parasitic nematodes.[136, 152]

Although the contents of the intestine of most animals are nearly devoid of oxygen there is a significant diffusion of oxygen from the mucosa into the lumen of the intestine; however, this oxygen is rapidly used by the microflora of the intestine resulting in a steep gradient from the mucosa to the centre of the lumen. Thus a small nematode, such as *Nippostrongylus*, which lives in close contact with the mucosa, will obtain a significant supply of oxygen and have an aerobic metabolism (under anaerobic conditions in a saline medium there is complete loss of motility within 6 hours followed by death in 6 to 24 hours[113]); while a nematode living in the contents of the intestine will have a predominantly anaerobic metabolism.

4: Osmoregulation and Excretion

Permeability of the Cuticle

The thickness and apparent complexity of nematode cuticle led to the general assumption that it was an inert, resistant and impermeable covering which protected the animal from the adverse effects of the environment. The cuticle of nematodes is, in fact, neither completely impermeable nor inert. It is permeable to certain ions, non-electrolytes and to water. The permeability has been found to vary with the nature of the environment in which the nematode normally lives. Marine nematodes, for example, have a cuticle which is more readily permeated than that of soil inhabiting forms. The cuticle of the free-living stage of *Agamermis* is relatively impermeable to *intra-vitam* stains but that of the parasitic stage, which lives in the body cavity of insects, is readily permeated by these stains. The female of the plant parasitic nematode *Heterodera* when mature becomes swollen with eggs, dies and forms a hard cyst. This cyst is at first white and soft, but owing to the activity of polyphenol oxidase in the cyst wall it darkens and hardens. Insect cuticle owes its impermeability to a layer of waxy material on the outside of the cuticle, but this layer is not present on the hardened cyst wall of *Heterodera* and, as would be expected, the cyst wall is permeable to water.[41]

The cuticle of most intestinal parasitic nematodes is usually less readily permeated than that of nematodes which live in the body fluids of other animals and of marine or soil inhabiting nematodes; however, it is permeable to water and to certain ions. Adult *Ascaris* and *Parascaris* when placed into a hypotonic or hypertonic solution swell or shrink accordingly, there is a change in weight and the freezing point of the body fluid changes showing that water can pass in and out of the body under the influence of osmotic forces (fig. 21). By ligaturing the nematodes at the mouth and anus it has been shown that this movement of water takes place across the cuticle and body wall as well as across the intestine.[63, 64]

Urea and potassium iodide, when introduced into an eviscerated *Ascaris*, rapidly pass through the body wall and cuticle and appear in the outer bathing fluid, but glucose does not pass through.[94] Glucose will pass through the cuticle, however, when the muscles and hypodermis are scraped away from it. The rate of penetration of non-electrolytes through the cuticle of *Ascaris* decreases with increasing molecular size.[63] Phosphate ions do not pass through the cuticle of living *Ascaris* but do after the nematode has been treated with a detergent; this suggests that the outer lipid layer is partly responsible for determining the permeability of the cuticle.[122]

Regulation of Body Fluids

Nature of the environment.[69, 152, 159] Nematodes are essentially aquatic organisms and must, therefore, be able to adapt themselves to changes in osmotic pressure of the environment to a greater or lesser extent. Marine species live in a constant environment and their tissues are probably isotonic with sea water so we should not expect an efficient osmoregulatory mechanism. Nematodes which live in the soil, however, are in an environment which is constantly changing. The pores between soil particles in arable soil contain air most of the time, with a thin film of moisture over the soil particles and ' lenses ' of moisture at the points of contact between particles (fig. 17*B*). Soil nematodes, covered by a thin film of moisture, are able to move freely under these conditions and are well supplied with oxygen. When the soil is flooded with rain, water percolates down into the pores in sudden surges as the surface tension at the pore necks is overcome and nematodes in the pore spaces are suddenly immersed in diluted soil moisture (fig. 17*A*). After rain the soil begins to drain and water is drawn out of the pores by suction. The size of the pores affects the amount of suction required to empty them, large pores needing less suction to empty them than small ones. Air returns to the soil pores and moisture is present as a film over the particles and as large lenses of moisture where the particles touch. As water is drawn out of the soil by evaporation and by the roots of plants the amount of soil moisture in the pores decreases even further (fig. 17*C*) leading to an increase in suction pressure and osmotic

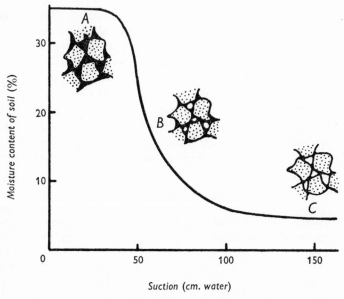

FIG. 17. The moisture characteristic of a hypothetical soil sample. *A*. Pores full of water; *B*. Pores emptying; *C*. Pores empty (after Winslow and Wallace, 1958. *Ann. appl. Biol*).

pressure. Eventually the soil may become so dry that the nematode is immobilized by surface tension forces and dehydrated by osmotic and suction forces. The ' moisture characteristic ' of a soil relates the percentage moisture content to the forces (suction pressure) holding water within the pores between soil particles (fig. 17). Thus water held in a soil at a suction of h cm. can be removed from the soil only if a greater suction than h cm. is applied to it (e.g. by evaporation, drainage and removal by plant roots). The soil moisture is usually a relatively dilute solution of electrolytes and organic components. The osmotic pressure rarely exceeds 2 atmospheres and is probably much less than the osmotic pressure of the body fluids of soil inhabiting nematodes. Thus nematodes in the soil will take up water from the environment. The suction pressure together with the osmotic pressure of the soil water gives the ' soil potential ' (pF). The pF of the environment is a most important ecological factor for soil inhabiting nematodes as gradients of soil potential exist throughout the soil profile and vary with changes in temperature, rainfall and the presence or absence of plants.[69, 152, 158] Thus, nematodes living in the soil are constantly being subjected to changes in soil potential and must have an efficient osmoregulatory mechanism if they are to survive.

Nematodes which live in the tissues of plants probably tolerate the same osmotic range as the host tissues. Nematodes which live as parasites in the tissues and body fluids of other animals are in a constant environment which is probably isotonic or hypertonic to the body fluids of the nematode. These species may therefore not have an efficient osmoregulatory mechanism or may actively excrete ions. Each region of the alimentary tract of animals is also a relatively stable environment, but fluctuations in the nature of the contents do occur. Nematodes which live in these environments will probably have limited powers of osmoregulation.

Changes in the ability to osmoregulate probably occur throughout the life cycle of animal parasitic nematodes which have free-living stages. The free-living larvae are essentially soil-dwelling species and have to cope with changes in soil potential. The majority probably live in a hypotonic environment and will take up water from the environment, although species which remain in the dung in which they were deposited may live in isotonic or hypertonic surroundings. The infective larvae thus have to be able to adapt to the hypotonic soil moisture, to a certain amount of drying when they are on the blades of grass (before being eaten), or on the soil surface (waiting to penetrate the skin of the host), and to the environment of the intestine or the tissues of the host.

Adaptation to the environment. Nematodes belonging to the genus *Rhabditis* occur in soil, decaying materials and in some animals as parasites. *Rhabditis terrestris* occurs as a resting larva in earthworms. When the earthworm dies the decaying tissues, and the accompanying bacteria, provide nourishment for the larva which then becomes adult. A few generations are spent on the earthworm until decomposition is complete when the larval stages migrate into the soil and enter another earthworm. Thus the nematode has to adapt itself osmotically to life in the soil, in the living earthworm and to variations in the composition of the bathing medium as the earthworm decays. Water can pass into

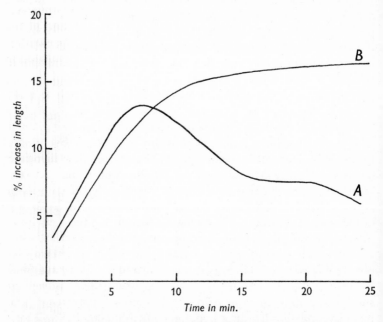

FIG. 18. The changes in length of *Rhabditis terrestris* when immersed in *A*, distilled water and *B*, M/100 KCN (after Stephenson[144]).

and out of the body of *Rhabditis terrestris* under the influence of osmotic forces. In isotonic saline there is no visible change in size, but in distilled water there is a rapid increase in length during the first 5 minutes together with a decrease in activity, probably because of increased hydrostatic pressure in the pseudocoelom. This is followed by a slow return to normal size and movement (fig. 18*A*). During the recovery period large amounts of fluid are expelled from the alimentary tract through the anus. That this recovery is due to an active method of osmoregulation is shown by the fact that recovery is inhibited by injury to the nematode or when potassium cyanide is present (fig. 18B). In hypertonic saline the body length decreases at first, but after a considerable time (20 hours) there is a partial return to normal length.[144]   In the above experiments the nematodes were transferred from one solution to another, thereby bringing about marked changes in body size.   Normally any changes in the environment will take place more slowly and the nematodes will be able to adapt to them more easily and without undue distortion.

Nematodes which inhabit brackish water are thought to be a mixture of marine species which can withstand a certain amount of dilution of sea water, and fresh water or soil-inhabiting species which can adapt themselves to increases in salinity. *Aphelenchoides parientinus*, for example, can tolerate wide ranges of salinity and is found in water varying in salt content from two to forty parts per thousand, while the soil inhabiting nematode *Panagrolaimus rigidus* is found in water varying in its salt content from one to forty parts per thousand. There is normally, however, a sharp dividing line between marine and fresh water species.[158]

Plant parasitic nematodes (such as *Heterodera* and *Meloidogyne*) have an efficient osmoregulatory mechanism.[40]   They can survive soil potentials (suction pressure plus osmotic pressure) up to the wilting point of plants and a pF which kills them by ex-osmosis will also bring about permanent wilting of the host.   Some can tolerate osmotic pressures up to 10 atmospheres, while others can withstand even higher pressures without being killed, although there is a cessation of movement under these conditions.[8, 152]   The plant parasitic nematode *Tylenchorhynchus* can withstand osmotic pressures equivalent to 1 M urea.   This is probably achieved by some active method of regulation for there is an increase in oxygen consumption with increasing osmotic pressure.[154]   In very dry soils, in which the osmotic pressure is high, the tendency of the suction forces plus the osmotic forces will be to desiccate the nematode and this may be of survival value to some species. *Ditylenchus dipsaci* can survive temperatures as low as −80° C. if water is removed from the body by previous immersion in concentrated salt solutions.

Parasites which inhabit the alimentary tract of insects are subjected to quite marked changes, including changes in osmotic pressure, when the host undergoes a moult.   In termites and in the cockroach, *Cryptocercus*, the flagellates in the alimentary tract are lost at each moult and reinfection of the insect occurs later.   In the cockroaches, *Periplaneta* and *Blatta*, the protozoa and the oxyurid nematodes which inhabit the hind-gut of these insects are

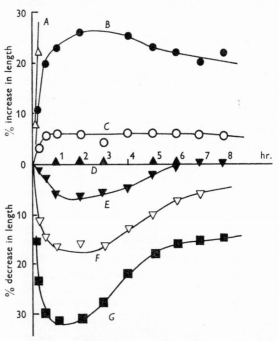

FIG. 19.   The percentage increase or decrease in length of females of *Hammerschmidtiella diesingi* in distilled water and in various concentrations of sodium chloride.   *A*. Distilled water;   *B*. 0·05 M NaCl;   *C*. 0·08 M NaCl;   *D*. 0·15 M NaCl;   *E*. 0·2 M NaCl;   *F*. 0·3 M NaCl;   *G*. 0·5 M NaCl (Lee[81]).

not lost during the moult.   As the insect completes the moult the nematodes escape from the cast peritrophic membrane in the posterior part of the hind-gut and move forward to their normal position, having survived the changes brought about by the moult. *Hammerschmidtiella*, which is one of these oxyurid nematodes parasitic in cockroaches, expands rapidly and bursts when placed in distilled water.   In hypotonic solutions the nematode increases in size but, unlike *Rhabditis*, there is no marked decrease in size, even after 24 hours (fig. 19*B*, *C*) which would suggest that the active method of osmoregulation found in *Rhabditis* is either absent or works much more slowly in *Hammerschmidtiella*.   In hypertonic solutions of sodium chloride or sea water the nematode recovers its normal length after a period of shrinkage (fig. 19*E*, *F*), which suggests that this nematode is able to adapt itself to increases in

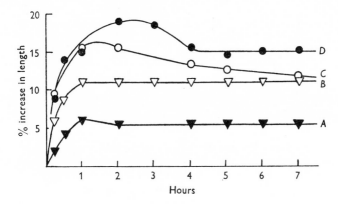

FIG. 20.   The percentage increase in length, in 0·15 M sodium chloride, of females of *Hammerschmidtiella diesingi* after 24 hr. in 0·2 M (*A*); 0·3 M (*B*); and 0·4 M NaCl (*C*); and of females of *H. diesingi* taken from moulting cockroaches (*D*).   0·15 M NaCl is normally isotonic (Lee[81]).

osmotic pressure much better than to decreases. This recovery is brought about by the uptake of ions, because nematodes which have recovered their normal length in a hypertonic medium increase in length as though they were in a hypotonic solution when placed in what is normally an isotonic one (fig. 20*A*, *B*, *C*). There is no decrease in length within 24 hours, which indicates that in this nematode the excretion of ions does not take place as readily as the uptake of ions.   In hypertonic sucrose solutions there is a steady decrease in length and no recovery of length, indicating that the cuticle of this nematode is relatively impermeable to the sugar molecules. *Hammerschmidtiella* taken from the hind-gut of a moulting cockroach, on the other hand, increase in length in what is normally an isotonic solution (fig. 20D), which suggests that the nematodes adapt themselves to an increase in the osmotic pressure of the hind-gut of the insect during moulting. Presumably, when the insect has completed the moult there will be a return to normal osmotic conditions within the hind-gut and in the nematode. This is likely to be a much more gradual process than the sudden transition which occurred in the experiments.[81]

Ascaris larvae (artificially hatched from the egg) can be maintained in Ringer's saline. If, however, the concentration of the solution drops below 0·7 per cent. there is a rapid shortening of the life of the larvae. There is only a gradual reduction above 1·1 per cent., the optimum concentration being 0·8 per cent. Ringer's solution.[50] This shows that Ascaris larvae, like Hammerschmidtiella, are more efficient at osmoregulation in hypertonic solutions than in hypotonic solutions. This is likely to be related to the hypertonic conditions encountered in the alimentary tract and blood system of the host.

Ascaris adults normally live in a hypertonic medium. When transferred to 30 per cent. sea water there is no appreciable change in weight, but the osmotic pressure of the body fluid is found to be slightly above that of the sea water, whereas when it is in the pig it is slightly below that of the intestinal contents. The chloride concentration of the intestinal fluid of the host varies from 34 to 102 m. mol. with an average of 66 m. mol. In the pseudocoelomic fluid of Ascaris it is fairly constant with an average value of 52 m. mol. The chloride of the pseudocoelomic fluid increases in dilute sea water but always remains below that of the external medium. This is brought about by an active process of chloride excretion against a concentration gradient. The excretory mechanism appears to be located in the body wall and not in the cuticle. Ascaris is also able to regulate the concentration of magnesium, calcium and potassium in the pseudocoelomic fluid.[64] Thus, although Ascaris has little in the way of a water regulating mechanism it has a system of ion regulation which is controlled mainly by the hypodermis. It is, however, dependent upon a relatively stable environment for the pseudocoelomic fluid concentration changes with changes in the medium (fig. 21C).

Angusticaecum, which lives in the intestine of the tortoise, is able to regulate its water content in a hypotonic medium, as it remains hypertonic to the environment when placed in tap water over a period of several days (conditions under which most animal parasitic nematodes burst). In sea water, however, there is an increase in the osmotic pressure of the pseudocoelomic fluid, showing that the nematode is unable to keep the pseudocoelomic fluid hypotonic to the medium. It can, however, tolerate changes

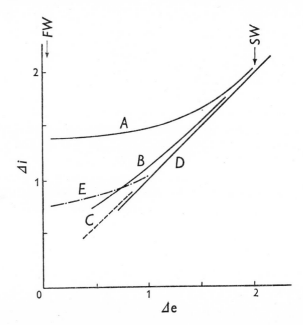

Fig. 21. Variation of internal with external osmotic pressure in *A. Carcinus*; *B. Nereis*; *C. Ascaris*; *D. Maia*; and *E. Angusticaecum* (Rogers[122]).

in the osmotic pressure of the pseudocoelomic fluid equivalent to 1 to 3·5 per cent. sodium chloride. The osmotic curve for *Angusticaecum* (fig. 21E) is similar to those of euryhaline invertebrates such as *Nereis diversicolor* (fig. 21B) which are able to live in a wide range of salinities.[104, 122]

Several other animal parasitic nematodes are able to adapt to varying osmotic pressures in their environment, as is shown by their ability to survive, sometimes for long periods, in solutions varying in sodium chloride content from 0·4 to 1·3 per cent.

Excretion of water and ions

Little is known about the excretion of water and ions by nematodes when they are in a hypotonic environment. *Rhabditis* expels some through the rectum and anus,[144] but there is little definite information whether or not the so-called excretory system plays any part in osmoregulation in nematodes.

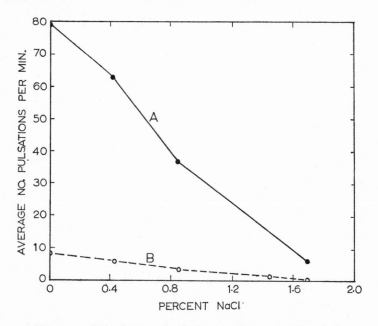

FIG. 22. Pulsation rates of the excretory ampulla of *A*, *Nippostrongylus brasiliensis* larvae and *B*, *Ancylostoma caninum* larvae in various concentrations of sodium chloride (after Weinstein, 1952. *Exp. Parasit.*).

Pulsations in parts of the excretory system of several nematodes have been described. These usually occur in the ampulla behind the excretory pore but in some species take place in the canals themselves. Experiments done on the third-stage larvae of *Nippostrongylus* and *Ancylostoma* show that the rate of pulsation of the excretory ampulla decreases with increase in concentration of the medium. In distilled water the ampulla of *Nippostrongylus* averages eighty pulsations per minute whereas the ampulla of

Ancylostoma averages eight.   In 1·7 per cent. sodium chloride the ampulla of *Nippostrongylus* averages only five pulsations per minute while pulsations stop in *Ancylostoma*[155] (fig. 22).

Calculations to determine the amount of fluid expelled through the excretory pore of these two nematode larvae show that a larva of *Nippostrongylus* (650 × 26 μ) excretes an amount of fluid equivalent to its body volume in 11 hours in distilled water, while a larva of *Ancylostoma* (680 × 22 μ) does so in 75 hours (table VIII).   There seems to be little or no relationship between

Table VIII

The amount of fluid expelled through the excretory pore per minute in μ^3 when third-stage larvae of Nippostrongylus *and* Ancylostoma *are kept in various concentrations of sodium chloride*[155]

Sodium chloride %	Nippostrongylus	Ancylostoma
0	530	57·5
0·42	419	40·5
0·85	246	24·1
1·45	100	8·5
1·70	34	3·8

osmoregulation and the rate of respiration in the larvae of *Nippostrongylus* because the level of oxygen uptake is always greatest in concentrations of sodium chloride between 0·75 and 1·0 per cent. at 37° C.[155]

In *Spironoura*, which is found in the alimentary tract of turtles, the excretory canals pulsate at their posterior extremities. The tube widens slowly and then contracts to such an extent that the lumen is obliterated.   The fluid contents move forward when the tube contracts and are carried along the length of the tube by a peristaltic wave, thus indicating that the wall of the excretory canal is contractile.[155]

In hypertonic media, enlargement of the amphids and phasmids occurs in some free-living species and it has been suggested that they play a part in the secretion of osmotically active material.[97] *Capillaria hepatica*, which lives in the liver tissue of its hosts, possesses glands which open to the exterior through a pore in the cuticle.   These are thought to have an osmotic or ion regulating function as the greatly folded membrane of the cell at the base of the pore (fig. 23) is reminiscent of glands which have an osmotic or ion regulating function in other animals.[159]

Summary. Although many nematodes have the ability to carry out osmoregulation, this ability is apparently more pronounced in

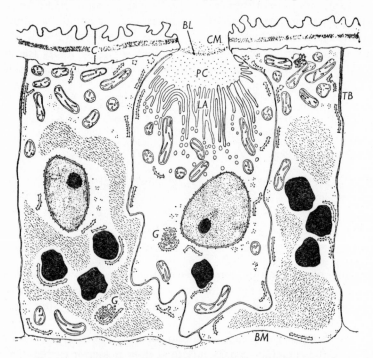

FIG. 23. Diagram of the hypodermal gland cell and non-glandular cells of the lateral and ventral hypodermal cords of *Capillaria hepatica*. The hypodermal gland cell is situated below a pore in the cuticle and is surrounded by non-glandular cells. These glands may have an ion-regulating function (Wright[160]).

> *BM*, basement membrane; *BL*, boundary layer; *C*, cuticle; *CM*, cap material; *G*, golgi; *LA*, lameller apparatus; *PC*, pore chamber; *TB*, terminal bar. Glycogen is represented by areas of irregular pattern.

species which normally encounter large fluctuations of osmotic concentration in their environment. In some nematodes the pseudocoelomic fluid concentration changes with the medium. These species have little or no osmoregulatory powers and, while the tissues of some are tolerant to a wide salinity range (euryhaline), others are limited to environments where there is little variation in the concentration of the medium.

Osmoregulation is, at least partially, brought about by the excretory system in some nematodes. In hypotonic solutions, in which water enters the nematode across the cuticle, excess fluid is collected by the excretory system and actively pumped out through the excretory pore. The differences in the rate of pumping and the amount of fluid expelled by different species of nematode living in the same environment (e.g. larvae of *Nippostrongylus* and of *Ancylostoma*) may be due to differences in the permeability of the cuticle or to the presence of other osmoregulatory mechanisms. The excretion of water or ions may also occur through the cuticle or the alimentary system thus lowering the osmotic concentration of the body fluids, as in *Rhabditis terrestris*. Other nematodes are able to excrete or take up ions from the environment when placed in hypotonic or hypertonic media.

Although many nematodes can adjust their internal osmotic pressure to a certain extent, in some quite marked changes take place in the internal concentration when the concentration of the medium changes. The tissues of nematodes must, therefore, be able to function in a range of concentration of electrolytes to a greater or lesser extent.

Excretion of nitrogenous and other waste substances

Nitrogenous excretory products. In nematodes the major nitrogenous waste product is ammonia. It is rapidly eliminated through the body wall and anus into the fluid environment where it is diluted to non-toxic levels.

When *Ascaris* swims freely in a saline medium it produces 69 per cent. of the total nitrogen excreted as ammonia and only 7 per cent. as urea. When the nematode is suspended in moist air in a U-tube, with only the head and tail in saline (fig. 24) the amount of ammonia produced falls to 27 per cent. of the total nitrogen excreted and the amount of urea produced increases to 52 per cent. (table VII). This indicates that under osmotic stress, when the amount of fluid taken in by the cuticle is lessened leading to an accumulation of ammonia in the tissues, this species can switch from ammonia excretion to the production of urea.[48, 129, 155] In this way the highly toxic ammonia, which is normally swept away in the fluid surrounding the worm, is removed as the much less toxic urea. It is not known whether this ability to switch from ammonia to urea excretion is present in other nematodes.

The amount and type of nitrogen excreted by several nematodes is given in table VII. In all cases the nematodes were existing solely on their food reserves.

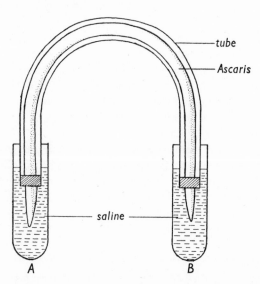

tube

Ascaris

saline

A   B

FIG. 24. *Ascaris* sealed in a U-tube containing moist air. The head and tail ends project from the tube and are immersed in saline (A) and (B). Most nitrogenous excretory products were detected in saline bathing the tail end and little in the saline bathing the head end which bears the excretory pore. In this situation *Ascaris* produces more urea and less ammonia than free-swimming individuals (after Savel[129]).

Free-living, plant and animal parasitic nematodes excrete amino acids but so do many other invertebrates. Large quantities of peptides are also excreted by some nematodes. Amines are not normally excreted by animals but several nematodes excrete a variety of these substances. These are all larvae of animal parasitic species.[155] The free-living nematode *Caenorhabditis* does not excrete amines.[127] The amines, which are strongly basic, may keep the environment near neutrality as some end-products of metabolism of nematodes are acidic. This may be of survival value for nematode larvae, such as *Ascaris*, which remain in the egg until eaten by the host.[122]

Other substances excreted. Nematodes parasitic in other animals excrete a variety of organic acids. These are mainly fatty acids which are end-products of carbohydrate metabolism (see chapter on metabolism) and are given in table IV.

The free-living nematode *Caenorhabditis* excretes only traces of the organic acids excreted by animal parasitic nematodes.[127]

Lactic acid is excreted by *Litomosoides* and *Dracunculus*[16, 21] but only traces of it are excreted by the other nematodes so far investigated. Some nematodes excrete a much more complex mixture of fatty acids than others (table IV).

Nematodes also excrete carbon dioxide, but some nematodes are able to fix metabolically produced carbon dioxide and use it in their metabolism.[48]

Excretion of waste substances. It has been demonstrated, in some nematodes at least, that the excretory system takes part in osmoregulation, but evidence demonstrating the excretion of waste products through the excretory system is meagre.

Droplets of a homogeneous material are extruded from the excretory pore of several species of *Rhabditis*. This material is acidic and rather insoluble in water.[26] A short length of some excretory product has been observed being expelled from the excretory pore of *Paraphelenchus*. This material remained in the water near the nematode, retaining its string-like form but swelling slightly.[56]

When infective-stage larvae of *Nippostrongylus* are placed in serum taken from rats which have become immune to this nematode, heavy precipitates form at the excretory pore. Similar findings have been demonstrated in several other larval nematodes. An antigen, which combines with the antibody in the host serum to give a precipitate, may be present in the fluid excreted from the excretory pore of the nematode. This, however, is rather indirect evidence of an excretory function for this system.[155]

More direct evidence has been obtained from the larvae of *Nippostrongylus* which excrete several primary aliphatic amines through the excretory pore. Ammonia and 1,2-dicarboxylic acids are also excreted through an unknown route. *Ascaris* larvae also excrete amines but the excretory route has not been determined.[48, 122, 155]

In some animals certain organs have been assigned an excretory function by injecting dyes into the body cavity and observing where the dye is excreted. Evidence from the use of dyes alone is, however, inconclusive. Living *Rhabditis*, stained with neutral red, exude drops of fluid containing this dye through the excretory pore, which suggests that in this case the dye is collected and excreted by the excretory system.[26]

Recently, fluorescent dyes have been used to detect excretory routes in several nematodes.[4] The nematodes were exposed to dye by injecting it into the host, by incubating the nematodes in a solution of the dye, and by injecting the dye into the body cavity of the nematode. Several animal parasitic nematodes (*Aspiculuris, Syphacia, Spironoura, Cosmocerca*) which have the H-type or the U-type excretory system (fig. 7), excreted several fluorescent dyes very quickly through the excretory pore after they had been ingested. The dye was not expelled continuously but was emitted in a series of fine jets, with pauses between one emission and the next, during which the excretory canals again filled with dye. These dyes are always excreted through the excretory canals and never through the cuticle although some dye is excreted through the anus. Eosin and phloxin do not pass from the body cavity into the canals and are not excreted. Some substance which absorbs and concentrates acridine orange, is passed out of the excretory pore of *Syphacia*. In animal parasitic nematodes which contain ventral 'excretory' glands as well as the lateral canal system (e.g. *Uncinaria* and *Oswaldocruzia*) (fig. 7) quite different results were obtained. Ingested fluorescent dyes concentrate in the cells of the intestine and do not pass into the pseudocoelom, while those injected into the pseudocoelom are not excreted and do not become concentrated in the excretory canals.[4, 155]

There is conflicting evidence about the function of the excretory system in adult *Ascaris*. Drops of fluid from the excretory pore of *Ascaris*, freshly removed from the intestine of the pig, contained about 0·02 per cent. urea. The concentration of urea decreased and disappeared when the nematodes were kept in saline for 24 hours.[26, 47]

Neutral red, Janus green or methylene blue, injected into the pseudocoelom of *Ascaris*, appear in the surrounding medium but no definite route of excretion has been determined. Urea injected into the pseudocoelom can be detected in the hypodermis and lateral lines but very little is found in the lumen of the excretory canals. These results have been interpreted as indicating the lack of an excretory function for the so-called excretory system in *Ascaris*. It was believed that excretion was through the cuticle as urea, potassium iodide, neutral red and methylene blue passed from the inside of eviscerated *Ascaris* into the bathing medium.[94] There is, however, no indication of the excretion of acid vital dyes through the cuticle of *Graphidium* when fed with the dyes in their natural habitat (intestine of the rabbit),[44] nor do fluorescent dyes pass out through the cuticle of *Aspiculuris*.[4]

Adult *Ascaris*, placed in a U-tube with the head end in saline and separated by a gap of air from saline bathing the rear end (fig. 24), transferred about 10 ml. of saline from the front tube to the rear tube in 24 hours. Little, or no, nitrogenous waste products were found in the front tube containing the mouth and excretory pore, whereas these waste products were easily detected in the saline in which the anus was immersed. This indicates that the excretion of nitrogenous waste products takes place through the anus rather than through the excretory pore in *Ascaris*.[129]

In some nematodes there may be a filtration of excretory products into the lateral excretory canals brought about by the high hydrostatic pressure of the pseudocoelomic fluid. If this is so, then a process of selective absorption of essential substances would be expected to occur somewhere along the length of the lateral canals but this has so far not been demonstrated.[61, 155] Esterase and amino peptidase have been detected in the walls of the excretory canals of *Ascaris* which shows that these canals are metabolically active.[82]

Thus, in certain nematodes the excretory system eliminates water, nitrogenous waste products and possibly ions and other substances from the nematode. In other nematodes, however, notably *Ascaris*, the function of the so-called excretory canals is still in doubt.

Several other functions for the excretory system of nematodes have been suggested. The ventral ' excretory ' glands may secrete an anticoagulant or digestive enzymes. These glands in *Nippostrongylus* contain large globules which have the same appearance under the electron microscope as large globules found in the pharyngeal glands and both types of globule contain large amounts of esterase, which would indicate that they have a similar function.[83] However, the ventral glands of *Strongylus* do not contain any proteolytic enzyme.[45] The excretory system may play a part in moulting in nematodes as discussed in chapter 5.

Much waste material is excreted through the anus and the cuticle of nematodes. One of the major excretory products of nematodes is ammonia and this is rapidly excreted through the cuticle and the anus. Carbon dioxide is almost certainly excreted through the cuticle. As mentioned earlier, urea is able to pass through the cuticle of *Ascaris* but it is not known if this is a normal excretory route. The excretory products of nematodes also include uric acid, amino acids, amines, peptides and fatty acids to a greater or lesser extent but the excretory route for these substances is unknown.

5: Hatching and Moulting

All nematodes have the same basic life cycle, which consists of an egg, four larval stages and an adult. Each larval stage is separated by a moult during which time the larva undergoes certain morphological changes, grows another cuticle and casts the old one. In some species the egg hatches immediately after embryonation is completed to release a first-stage larva. Other species hatch after the first or the second moult has taken place inside the egg. This delayed hatching of the egg is thought to be an adaptation to parasitism.

Nematodes grow between moults and the mature worm may continue to grow after the final moult has taken place (*Ascaris*, *Parascaris*, *Heterodera* females and *Meloidogyne* females).

The Nematode Egg

The eggs of nematodes vary in shape and structure (fig. 25) but are essentially ovoid and have three main layers; (1) a thin inner membrane, (2) a thick chitinous layer and (3) an outer protein layer (fig. 25*E*).

The inner membrane was, until recently, thought to be composed of lipid but it is now known that in *Ascaris* it is composed of esterified glycosides with the solubility characteristics of lipids. Whether or not this membrane in the eggs of other nematodes is also composed of glycosides is not known. The relative impermeability of nematode eggs to water soluble substances has been attributed to the inner membrane.

The chitinous layer of the egg shell, which is secreted by the

81

egg and not by the female reproductive tract, is the only structure in nematodes which possesses chitin. It also contains protein. In many species this layer is not present at one or both ends of the egg, and so forms an operculum (fig. 25 *B*, *D*).

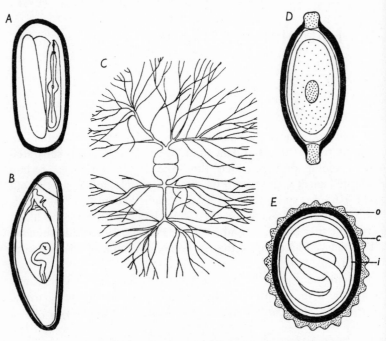

FIG. 25. Eggs of some nematodes. *A*. Tylenchid egg containing larva; *B*. Oxyuroid egg (*Leidynema*) containing a contracted ' resting ' larva; *C*. *Mermis* egg; *D*. *Trichuris* egg; *E*. *Ascaris* egg containing larva. (*A* and *C* after Jones, 1959. *Plant Nematology*, H.M.S.O.)

c, chitinous layer ; *i*, inner layer ; *o*, outer layer

The outer layer of the egg is not present in all species. It is secreted by the wall of the uterus and is often characteristically marked or drawn out into processes (fig. 25*C*).

Hatching

Hatching to release free-living stages

In free-living nematodes the eggs usually release a first-stage larva, and hatching is controlled partly by external factors such as temperature and moisture, and partly by the unhatched larva itself. Thus, for hatching to occur the egg must have developed sufficiently and the external conditions must be suitable. This mechanism ensures that the egg does not hatch during periods of adverse conditions. Before hatching, the egg becomes permeable to water, possibly due to enzymes released by the larva attacking the inner layer of the shell.

Many animal parasitic nematodes which have free-living larvae (*Ancylostoma*, *Nippostrongylus*, *Trichostrongylus*) are similar to the free-living nematodes in that they hatch when the first-stage larva is fully formed, but only under suitable environmental conditions. During hatching of the eggs of the animal parasitic nematode *Trichostrongylus retortaeformis* the larva moves freely inside the egg and this movement, together with the action of enzymes, breaks down the innermost layer of the egg shell. Once this layer is broken down the egg shell becomes permeable and the larva takes up water, which results in an increase in hydrostatic pressure in the pseudocoelom. As a result the larva becomes relatively immobile, enlarges and exerts pressure on the shell, eventually rupturing it. It is also possible that the outer layers of the shell are weakened by secretions from the larva.[157] There is therefore no fundamental difference in the method of hatching of the eggs of free-living nematodes and of animal parasitic nematodes which have free-living larvae.

Eggs of the sheep parasite, *Nematodirus*, also hatch to release free-living larvae but the eggs must be chilled before they will hatch. This is similar to the diapause of some insects and has obvious survival value for the nematode. The adult does not usually survive in the sheep from one season to another but the eggs are able to withstand winter conditions and do not hatch unless subjected to a period of low temperatures. The eggs, therefore, do not hatch in the autumn when they are deposited but only in the spring when uninfected lambs begin to graze on the infected pasture.

Many animal parasitic nematodes and a few plant parasitic nematodes, which hatch from the egg as second- or third-stage larvae, require a specific stimulus to induce them to hatch. This seems to be an adaptation to parasitism, as quiescent larvae enclosed in the egg survive better and for longer periods than hatched larvae which must quickly find a suitable host.   The stimulus usually comes from a specific host or from closely related hosts.

Hatching of plant parasitic nematodes

The eggs of most plant parasitic nematodes hatch at a certain stage of development under suitable environmental conditions. In some species of *Heterodera*, however, hatching is markedly stimulated by substances from the host and *H. göttingiana* will hatch only in soil containing host roots.   The nature of the stimulus, which increases the rate of hatch, varies from species to species and is usually contained in a diffusate from the roots of the host but may sometimes be provided by the environment.   The stimulus from the plant is not essential for the hatching of most species of *Heterodera* as there is often an appreciable rate of hatch even in distilled water.   A proportion of the eggs (30 to 50 per cent.) in soil hatch in the spring without any plant stimulus, although soil organisms may be partly responsible for this, but in the presence of host roots up to 90 per cent. of the eggs may hatch. The whole process of hatching is sensitive to external conditions and is very variable and difficult to measure.[136]

The ' hatching factor ' responsible for the hatching of several plant parasitic nematodes has been studied intensively because of its potential importance as a means of controlling the nematodes, but the exact nature of the stimulus, which varies markedly from species to species, is still a controversial subject.   The stimulus from the host which initiates hatching of the eggs of *Heterodera rostochiensis* is a single factor.   The ' hatching factor ' is acid, unsaturated and is very unstable.   It may have the approximate formula $C_{11}H_{16}O_4$[105] and may contain an unsaturated lactone ring. The natural stimulus which increases the rate of hatch of *Heterodera schachtii* eggs is apparently a combination of factors. Eggs of *H. schachtii*, which has a wide host range, are stimulated to hatch by a variety of substances.   While some of the stimuli cause only a slight hatch, others are quite as effective as root diffusates from the host.[136, 152]

The eggs of *Heterodera* are enclosed in a cyst, formed by the hardening of the cuticle of the dead female, while in some other nematodes a jelly-like mass surrounds the eggs (*Meloidogyne, Tylenchulus, Rotylenchulus* and several species of *Heterodera*).

This complicates hatching as the stimulus has to penetrate the cyst or jelly before it reaches the eggs. The jelly-like mass containing the eggs differs in origin in different genera. In *Meloidogyne* it is secreted by the rectal glands, in *Tylenchulus* it comes from the excretory system, and in those species of *Heterodera* which secrete a jelly-like mass it may come from the walls of the uterus.[87] The cyst of *Heterodera* may act as an ecological unit and hatching of eggs in the cyst may be limited by an interaction of factors inside the cyst.[152]

Maximum emergence of larvae from the eggs and cyst of *Heterodera schachtii* takes place at dilutions of urea, sodium chloride and sucrose corresponding to an osmotic pressure of $0·48$ atmospheres. At weaker dilutions or in more concentrated solutions the numbers of larvae emerging are markedly reduced, but the hatch may be lessened because of indirect adverse osmotic effects and not because the osmotic pressure is directly stimulating hatching.[136, 152]

Oxygen is an important factor in the hatching of plant parasitic nematodes and the rate of emergence decreases with decreasing oxygen tension of the environment. The rate of emergence depends on the structure of the surrounding soil, porosity (fig. 26), microflora, microfauna and the amount of organic matter present, all of which influence the oxygen relations of the soil. The amount of water present around the cyst regulates the amount of oxygen reaching the eggs and hence regulates the rate of hatch. In water-logged soils the rate of hatch is very low but hatching increases when air appears in the soil pore spaces. Maximum emergence of larvae occurs in the soil when most of the pore spaces are almost emptied of water (fig. 17B). Eggs fail to hatch in soils with a very low moisture content (high suction pressure) (fig. 26)—this may be caused by a lack of oxygen within the cyst, the oxygen having been used by a few hatched larvae which are unable to escape from the cyst because of the lack of moisture, but is more probably because the lack of moisture may directly inhibit hatching.[136, 152]

Hatching of these plant parasitic nematodes is preceded by active movements of the larva within the egg. The egg shell also becomes flexible after secretions are produced by the larva. The mouth stylet may assist in the hatching process, being used to puncture the egg shell. *Meloidogyne* makes seventy to ninety thrusts a minute against the shell and repeatedly returns to the same spot on the shell. *Neotylenchus*, however, hatches in a similar manner to the animal parasitic nematode *Trichostrongylus*, as described earlier, and apparently does not use the stylet when hatching.[145, 152]

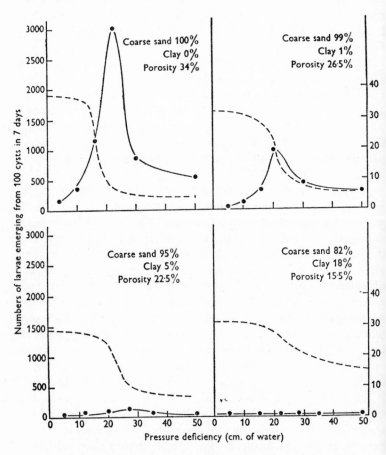

Fig. 26. The relationship between emergence of *Heterodera schachtii* larvae and the amount of moisture present in mixtures of sand and clay. The dotted line represents the moisture characteristic; the continuous line represents the numbers of larvae emerging (Wallace, 1956. *Ann. appl. Biol.*).

Carbon dioxide apparently inhibits hatching of *Heterodera* eggs as no larvae emerge from cysts in root diffusate saturated with carbon dioxide.[152]

Hatching in the animal host

More is known about the hatching of eggs of animal parasitic nematodes than about those of free-living and plant parasitic species. The hatching of eggs of animal parasitic nematodes with a free-living stage in the life cycle is essentially the same as for most free-living species. The stimuli for eggs, which have to be eaten by a definite host before they will hatch, are usually supplied by the host. While plant parasitic nematodes are stimulated to hatch from the egg at a distance from the host—by the environment or by diffusates from the host plant—the infective egg of animal parasitic nematodes is stimulated to hatch after it has been ingested by the host.

Carbon dioxide is the most important factor initiating the hatch of *Ascaris* eggs.[121] The presence of reducing agents, the addition of salts, variations in pH and temperature affect the rate of hatch, but only when dissolved gaseous carbon dioxide and undissociated carbonic acid [H_2CO_3] are present. Large concentrations of carbon dioxide inhibit hatching.[121] Hatching of *Ascaris* eggs takes place over a wide range of concentrations of [H_2CO_3] and dissolved gaseous carbon dioxide at pH 6, but as the pH is raised to 8 the effective range of concentrations which stimulate hatching is lessened although the actual numbers of eggs hatching increases. Thus at pH 7·3 more eggs hatch at the optimum concentration of [H_2CO_3] than at the optimum concentration at pH 6, but the effective range of concentrations at pH 7·3 is $0·2 \times 10^{-3}$ M to $0·3 \times 10^{-3}$ M while at pH 6 the optimum concentration of [H_2CO_3] is 1×10^{-3} M to 2×10^{-3} M. The eggs of *Toxocara* and *Ascaridia* behave similarly.[122, 124]

Different regions of the alimentary tract of animals have different pH values, different concentrations of undissociated carbonic acid and dissolved gaseous carbon dioxide, and varying oxidation-reduction potentials and oxygen pressures. Thus the eggs of animal parasitic nematodes are stimulated to hatch only in the correct host and in that region of the alimentary canal where the combination of factors necessary to initiate hatching is found.

The stimulus from the host causes the larva inside the egg to begin the hatching process. This has been demonstrated by exposing the eggs of *Ascaris* to brief stimulation, after which the eggs continue to hatch.[121] The hatching larva produces a chitinase, a lipase and a protease.[121] The enzymes first dissolve an area of the outer part of the shell. The larva then protrudes from the hole, still encased in the inner layer which eventually ruptures or is

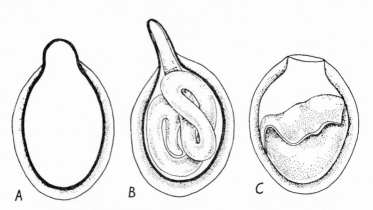

FIG. 27.   Stages in hatching of *Ascaris* eggs.   *A*. The stimulus from the host has produced an internal response from the larva (not shown) resulting in digestion of a hole in the outer layers of the shell through which the flexible inner layer protrudes. *B*. The larva has inserted its head into the blister and appears to be stretching it forcibly. *C*. The inner layer has ruptured and the larva has escaped from the egg leaving the inner layer of the shell collapsed inside the egg (Fairbairn[49]).

digested (fig. 27). This inner layer of the shell is usually very impermeable but during hatching the enzymes which attack the outer layers of the shell must pass through this layer. One of the first effects of the stimulus on the egg must be to alter the permeability of the inner layer, as occurs in the eggs of free-living and animal parasitic nematodes which release a free-living larva. This change in permeability after stimulation has been demonstrated by measuring the amount of trehalose which appears in the medium bathing the eggs. Trehalose, which is present in large amounts in the fluid bathing the larva inside the egg shell, does not leak from unstimulated eggs, but does escape from eggs which have been stimulated.[49]

Moulting

Very little is known about the process of moulting in nematodes. It happens at intervals during the development of the nematode and is probably controlled by internal secretions, as in insects.

Moulting occurs in three main steps:—1. the formation of the new cuticle; 2. the loosening of the old cuticle; 3. the rupture and ecdysis of the old cuticle with the ensuing escape of the larva.

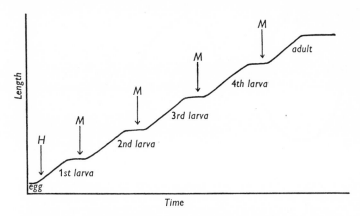

FIG. 28.   The life cycle of a hypothetical nematode showing the periods of growth and lethargus.

H, egg hatches ;   *M*, moult occurs

The larvae are able to grow between moults and in many species there is a lag period, in which little growth occurs, before, during and after moulting (fig. 28).   The loosening of the old cuticle is probably brought about by a moulting fluid secreted by the larva.   As in insects, part of the dissolved cuticle may be reabsorbed, only the outer layer being shed.

During moulting in plant parasitic nematodes, the basal part of the stylet dissolves and the head is disengaged from the anterior part of the old stylet which remains attached to the cast cuticle. In some species the new stylet develops in the wall of the pharynx and then migrates to its position in the buccal cavity.   Upon completion of the moult the nematode escapes from the old cuticle by rupturing it with the stylet or by abrasion against soil particles.[152]

Most research on the loosening and casting of the old cuticle has been done on animal parasitic nematodes which have two free-living larval stages before the third infective stage (*Haemonchus, Trichostrongylus, Ostertagia*). This third-stage larva is completely enclosed in a protective sheath formed by the old cuticle of the second-stage larva which has not been cast off. The first- and second-stage larvae feed on bacteria in the faeces of the host but the third-stage larva does not feed; it is entirely enclosed by the cuticle of the second-stage larva in which even the old anal, oral and excretory pore orifices are closed. These ensheathed larvae are, in many ways, similar to the infective eggs of *Ascaris* in that they have a lowered rate of metabolism. Ingestion by a suitable host provides the stimulus for exsheathment and the completion of development. The process of exsheathment of these third-stage larvae is rather similar to the hatching of *Ascaris* eggs and the stimuli (carbon dioxide and other co-factors) are essentially the same. The contents of the alimentary tract of the host provide the stimuli which cause the larva to secrete an exsheathing fluid. This exsheathing fluid then attacks the sheath at a specific area near the anterior end causing a weak ring to appear around the sheath. The inner layer of the cuticle is dissolved by the exsheathing fluid and the outer layer is ruptured by movements of the enclosed larva (fig. 29). A cap of cuticle becomes detached and the larva escapes by wriggling free.[78]

Host enzymes are not responsible for the digestion of the sheath of *Haemonchus contortus* larvae and *Trichostrongylus axei* larvae. This was shown by enclosing the infective larvae in a cellophane sac and placing them in the rumen of sheep, where exsheathment normally occurs. The digestive enzymes of the host are unable to pass through cellophane and yet the larvae were stimulated to exsheath. Digestion of the sheath, with the release of the cuticular cap, occurred as in normal exsheathment.[124] Fluid taken from exsheathing larvae also brings about exsheathment of unstimulated larvae. The infective larvae of *Dictyocaulus viviparus* and *Trichostrongylus colubriformis* require pepsin from the host as well as carbon dioxide to bring about exsheathment.[137]

The whole process of exsheathment takes about 3 hours but the action of the stimulus is completed after 15 minutes. Larvae exposed to the stimulus from the host for 15 to 30 minutes and then removed from the stimulus, continue to exsheath in water at 38° C.[124]

Exsheathment of larvae of *Trichostrongylus axei* and of *Haemonchus* increases as the concentration of $[H_2CO_3]$ increases. These larvae require much higher concentrations than the larvae

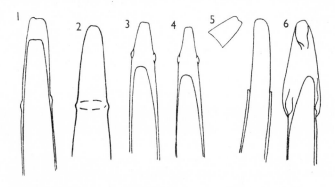

FIG. 29.   Stages in exsheathment of infective larvae of trichostrongyle nematodes (*Trichostrongylus*, *Haemonchus* and *Ostertagia*). The sheath, enclosing the third-stage larva, swells locally near the anterior end (1, 2) and separates into two layers (3). The inner layers are digested (4) resulting in an area of weakness in the form of a ring. The tip of the sheath breaks off in the form of a cap (5) and the larva wriggles free. (6) A larva with unusual distension of the cap area and rupture of the inner layer of the sheath (Lapage[78]).

of other nematodes examined and are the only species in which the process is not inhibited by a high concentration of carbon dioxide. *Trichostrongylus axei* normally exsheaths in the rumen. Maximum exsheathment takes place between pH 7 and 8 when the amounts of dissolved gaseous carbon dioxide and undissociated carbonic acid present are above 0.5×10^{-3} M. Larvae of *Haemonchus* need about three times the concentration of undissociated carbonic acid required by *T. axei* to bring about exsheathment but the pH range is the same. This may explain why *Haemonchus* has such a narrow host range, for the high concentration of $[H_2CO_3]$ necessary for exsheathment is found in relatively few situations other than the rumen of ruminants. *Trichostrongylus colubriformis* normally exsheaths in the acid stomach and it requires 5×10^{-3} M $[H_2CO_3]$ at pH 4 to bring about maximum exsheathment.[121] Thus the adult worm can be expected to settle in the region of the alimentary tract slightly posterior to that where conditions necessary for exsheathment are found.

The processes involved in exsheathment are probably not the same as those which initiate moulting in nematodes. In ensheathed nematodes moulting has already taken place and exsheathment only brings about the release of the larva from the sheath.

In insects certain neurosecretory cells control moulting but there is no evidence that this is so in nematodes, nor is there any direct evidence to show where the moulting or exsheathing fluid is released. Experiments with ligated *Trichostrongylus* larvae and with larvae exposed in various regions to a 60 μ wide beam of ultra-violet light, have shown that the entire mechanism which initiates exsheathment and causes the accumulation and subsequent release of exsheathing fluid lies between the base of the pharynx and the region of the nerve ring or the excretory pore.[124] The exsheathing fluid is probably released through the excretory pore of the ensheathed larva into the space between the cuticle and the sheath. It attacks the sheath at a definite area of weakness. The stimulus does not cause the larva to manufacture exsheathing fluid, as the fluid is already there, but it causes the larva to release the fluid. The nature and exact site of the cells which manufacture and store the exsheathing fluid are not known.[124] In those free-living nematodes in which moulting is initiated by the larva when external conditions are suitable, and probably also in all other nematodes, it is possible that neurosecretory cells release hormones, which activate the tissues bringing about changes in structure and activity, and stimulate the production of moulting fluid. Unfortunately, there is no experimental evidence that this is so. There is in addition a probable relationship between size and the onset of moulting in most nematodes.

Parasitic nematodes which require a specific stimulus from the plant or animal host to initiate hatching or exsheathment may have lost part of the internal mechanism which regulates these processes so that the parasite is dependent on the host to replace it.[122]

Growth

The growth of nematodes is a discontinuous process in that they undergo four moults during development (fig. 28). Unlike insects which increase in size during and immediately after a moult, nematodes increase in size between moults and, in the majority of cases, they stop growing (lethargus) just before, during and just after a moult (fig. 28). The duration of the lethargus differs in different species. *Ascaridia galli*, in the chicken, continues to grow without any sign of a lethargus.[1] Other animal parasitic nematodes (*Cooperia* and *Ancylostoma*) have definite periods of growth, lethargus, ecdysis and growth,[141] while the plant parasitic nematode *Meloidogyne* has a prolonged lethargus during which three successive moults occur.[5] In some instances the larva may be shorter after moulting than before. This occurs in the plant parasitic nematode *Heterodera* and in the animal parasite *Haemonchus*. The total mass of the larva may not change however, as the shortened larva may be thicker.

The fact that nematodes grow continuously, except during moulting or when they have reached the maximum size, presents certain problems as to how this occurs. The cuticle acts as an exoskeleton and the body wall musculature is attached to it. It is also a complicated structure consisting of two or three layers which, in large nematodes like *Ascaris*, may be further sub-divided into several more layers. To allow continuous growth to occur the cuticle must either stretch or increase in size. While this has not been studied in larvae, it is well known that *Ascaris* grows in size after the final moult, eventually reaching a length of approximately 20 cm., and that the cuticle increases in thickness as the nematode increases in length. This means, in *Ascaris* at least, that more cuticle is being laid down as the nematode grows. No mitochondria or nuclei are present in the cuticle of nematodes, but there is some evidence that it is metabolically active as enzymes have been detected in various layers of the cuticle of *Ascaris* and *Nippostrongylus*.[48, 83] It is not known whether these enzymes help to produce new layers of the cuticle or serve some other purpose.

Growth of the internal structure of the nematode is apparently less complicated than growth of the cuticle, as the cell number appears to remain constant and growth occurs by an increase in size or change of shape of a structure. It is possible that during development cells are destroyed and replaced to form new structures (e.g. change in shape of the pharynx from one larval stage to another) but there is no information on this. The cells of the intestine of some nematodes increase in number in the adult stage and regeneration of intestinal cells will probably occur after holocrine secretion has taken place.

Adaptations of the Life Cycle

Free-living nematodes have, on the whole, a straightforward life cycle with no prolonged lethargus at any stage. Some free-living nematodes, however, inhabit temporary environments, such as dung, or environments which are subject to drying, such as moss tussocks and the upper layer of soil of arable land, and these species usually have a stage of the life cycle which is adapted either for survival or for dispersal. The dispersal stage in the life cycle of many dung-inhabiting species is the third-stage larva. As well as having structural and metabolic adaptations this stage also has behaviour patterns which aid dispersal. *Rhabditis dubia* lives in cow dung and spends several generations in one deposit of dung. When the environment becomes unsuitable, however, resistant third-stage larvae (dauer larvae) are produced and these migrate to the top of the dung where they become attached to flies which visit the dung. When the fly alights on freshly deposited dung the larvae leave the fly and colonize the fresh dung. These third-stage larvae are a resting stage which is adapted to withstand desiccation during the flight of the fly.[9] They are enclosed within two cuticles, do not feed, and in many ways resemble the ensheathed infective larvae of certain animal parasitic nematodes.

Rhabditis coarctata is similar to *Rhabditis dubia* in that it inhabits dung, but the third-stage larva of this species is always a dauer larva and must be transferred to fresh dung to complete the moult.[9] Some stimulus is probably required from fresh dung to bring about exsheathment and further development of *R. coarctata*, rather as ensheathed larvae of *Haemonchus* and *Trichostrongylus* require a stimulus from the contents of the rumen of sheep.

The larval stages of plant parasitic nematodes which live in the

surface layers of arable soil, or nematodes which live in environments which periodically dry out, such as in moss tussocks or
small pools, are usually able to withstand a certain amount of
drying. One of the best known examples is *Anguina tritici* which
causes galls in wheat grains. The larvae of this nematode can
remain in a dry state for several years. The environmental
conditions which bring about maximum expulsion and movement
of larvae from the gall occur only when conditions in the soil are
suitable for seedling germination. Thus the infective larvae will
be released only when new hosts are germinating. The larvae of
many common plant parasitic nematodes can tolerate dry conditions in the soil for several days if the air is humid.

Nematodes which parasitize insects usually have varied, and
often complex life cycles. Some infect the insect when the
infective egg is eaten by the insect, but others penetrate the insect
(usually the insect larva) as a larval stage and live in the haemocoele
of the insect where they reproduce and cause the death of the host.
Parasitylenchulus diplogenus has a life cycle of two generations
which is intimately connected with the life cycle of the insect host.
The inseminated female penetrates the cuticle and enters the
haemocoele of the larval host, where it develops, and when the host
pupates it lays eggs in the haemocoele of the host. These eggs
hatch to release active larvae which develop in the haemocoele of
the adult insect, where they mate and lay eggs. Larvae hatching
from these eggs then leave the female host by descending the
oviduct to the vulva and are deposited on or near material in which
the host larvae occur, or pass out with the eggs of the host.
Copulation takes place at this stage and the inseminated females
enter another larval host.[156]

Adaptations of the life cycle of nematodes, which have become
parasitic in vertebrates, are often very complex. Some species
hatch from the egg to release free-living stages which feed upon
bacteria in the faeces and the soil. In some the third-stage larva
is a resting, infective stage which must be eaten by the host for the
life cycle to be completed. The host supplies the necessary
stimulus to continue development, and the position in the alimentary tract where this occurs partly determines the environment
of the adult (*Haemonchus*, *Trichostrongylus*). In other species,
stimulation of the infective larva takes place in the alimentary

canal but the larva then migrates through the tissues and vessels of the host to the final location (*Dictyocaulus*).

Many nematodes enter the final host through the skin, usually as the third-stage larva, after an initial free-living existence and migrate through the tissues and vessels of the host before settling in the tissue or organ where they develop to maturity (*Ancylostoma, Necator, Nippostrongylus*). Some have no free-living existence but develop to the infective stage in the egg. When ingested by a suitable host the egg is stimulated to hatch by factors, often very selective, in the alimentary canal of the host. The released larva may then complete development in the alimentary canal (*Oxyuris*), or penetrate the wall of the alimentary canal, migrate through the tissues and blood vessels to the lungs and then return to the alimentary tract down the oesophagus (*Ascaris*).

Pre-larvae (microfilariae) of the filarial nematodes are transferred from the blood stream of an infected host to that of another host by such biting insects as mosquitoes. The microfilariae undergo a period of development in the insect before migrating to the mouth parts of the insect. When the insect bites a host the nematodes are released on to the skin of the host and enter the puncture made by the insect. They are not injected into the host directly. The microfilariae of one species are transmitted by a night feeding mosquito and it is interesting that the microfilariae circulating in the blood of the host only come to the cutaneous blood vessels during the night, but if the host sleeps during the day and works at night the microfilariae come into the cutaneous blood vessels during the day. Another species, which is transmitted by day feeding insects, has microfilariae which migrate into the cutaneous blood vessels during the day and disappear from these during the night. This has obvious advantages for the transmission of the nematode.

There are many other examples of adaptations of the life cycle to parasitism and these are described in various textbooks of parasitology.

6 : The Nervous System and Sense Organs

The Nervous System

Structure

The nervous system of nematodes is basically the same in all genera. A nerve ring, made up of fibres with a few ganglion cells, encircles the pharynx. A few species of nematodes are known to have two nerve rings. Associated with this nerve ring are a small dorsal ganglion, two or more lateral ganglia and a ventral ganglion, which is sometimes divided into two. Six, or sometimes eight, longitudinal nerves run posteriorly from these ganglia. Six nerves pass forwards from the nerve ring and supply the lips and their associated sense organs. The ventral nerve (which runs in the ventral cord) is the largest nerve and is a paired structure for some of its length; the dorsal nerve (which originates from the dorsal ganglion and runs in the dorsal cord) is much smaller. A small nerve which arises from the lateral ganglion runs posteriorly along each lateral cord, and in many nematodes subdorsal and subventral nerves are also present. The dorsal and lateral nerves have a few ganglia associated with them but the ventral nerve has several ganglia along its length. Several commissures, running in the hypodermis, connect the longitudinal nerves at regular intervals along the length of the nematode. In the region of the nerve ring in some free-living and plant parasitic nematodes, a clear area in the hypodermis on the ventral side is a ventro-lateral commissure of the nervous system and is called the hemizonid.[56] A pair of nerves runs forwards from the lateral ganglia to the amphids at the anterior end of the nematode. The dorsal nerve is said to be chiefly motor and the lateral nerves mainly sensory in function. The submedian and the ventral nerves are partly motor and partly sensory.

There is a system of three nerves in the pharynx, one in each sector, which are connected with one another by commissures and also with the nerve ring.

Nematodes are unique in that the muscle cells of the body are innervated by processes which pass from the muscle to the nerve (fig. 2) and not, as in other animals, by nerves sending fibres to the muscle. It is claimed that the nerve-muscle junction is similar to those found in other animals,[62] although more recent work suggests that the neuro-muscular junctions in *Ascaris* are established between the nerve cord and a syncytium which branches off into the muscles.[35, 36]

Transmission along nerves

Almost nothing is known about the processes involved in the conduction of an impulse along nerves in nematodes. In *Ascaris*, however, it is known that the pseudocoelomic fluid contains more sodium (129 m. moles/litre) as compared with potassium ions (25 m. moles/litre). It is thus possible to speculate that the nerve axons in this species function as in other animals, in which the action current arises from an influx of external sodium ions; the resting potential depending on the excess of potassium ions in the axoplasm relative to the external medium.[108]

Neuro-muscular and synaptic transmission

General remarks. In most animals chemical substances are thought to be responsible for the transmission of impulses between nerve cells and across the neuro-muscular junction.[71, 108] The most widely distributed chemical transmitter substance is acetylcholine, the nerves secreting acetylcholine at synapses and muscle end-plates being referred to as cholinergic nerves. Acetylcholine is destroyed in the tissues by cholinesterase.

In vertebrates adrenalin and noradrenalin are liberated by the sympathetic nerves which supply the heart and smooth muscle. Sympathin (a mixture of adrenalin and noradrenalin) appears to be a transmitter substance in certain sympathetic neurons, but although it is present in some invertebrates its function in them has not been ascertained.

Serotonin (5-hydroxytryptamine) is found in the nervous system and tissues of many animals. It is a potent stimulant for the nerve net of coelenterates and may act as a cardio-accelerating substance in some invertebrates. The functions of serotonin are still largely undetermined.

Amine oxidases act on adrenaline and serotonin and their distribution parallels the distribution of these transmitter substances. For further reading see Katz[71] and Prosser and Brown.[108]

In nematodes. Acetylcholine is apparently involved in nervous transmission in nematodes. Acetylcholine-like substances have been isolated from three species. The 'head' of *Ascaris*, which contains the nerve ring and ganglia, has fifteen times more acetylcholine-like activity than strips from the body wall.[90]

Cholinesterase has been demonstrated in homogenates of *Ascaris* and *Litomosoides*, although only in small amounts as compared with that present in *Schistosoma* (Trematoda).[17] Cholinesterase is present in the innervation processes and in the sheath of the muscle of *Ascaris* and is abundant at the nerve-muscle junction and at the junction between the muscles and the nerve ring (fig. 30). The pharyngeal nerves and the nerves supplying sensory terminals in *Ascaris* are cholinergic and the enzyme appears to be located on the outside of the nerve. Parts of the nervous system of *Ascaris* contain an esterase enzyme which is not inhibited by cholinesterase inhibitors and it is possible that there may be some transmitter substance, other than acetylcholine, involved in nerve transmission.[82] The nervous system of some insects contains at least three types of esterase: an acetylcholinesterase inhibited by eserine and some organic phosphates, an aliphatic esterase inhibited by organic phosphates but not by eserine, and an aromatic esterase inhibited by neither. It is also suspected that serotonin is a transmitter substance in *Ascaris*.

Cholinesterase is present in several free-living and plant parasitic nematodes.[125] The most active areas in these nematodes are the nerve ring and associated ganglia, the amphids, phasmids and other sense organs. Plants of the Mary Washington variety of Asparagus are resistant to attack by plant parasitic nematodes. The plant releases a glycoside which not only kills nematodes (death is preceded by abnormal twitching and paralysis which indicates that the nervous system is affected) but also inhibits human plasma cholinesterase. The glycoside presumably inhibits the cholinesterases of the nematodes which approach the roots of the plant and thus paralyses them.[125]   Organic phosphate

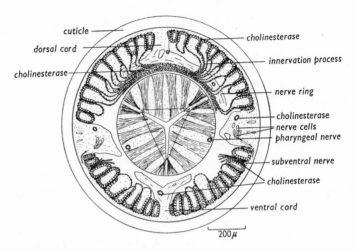

FIG. 30. Transverse section through *Ascaris* in the region of the nerve ring showing the distribution of cholinesterase (Lee[82]).

insecticides, which are known to inhibit cholinesterase in other animals, are toxic to nematodes and it is presumed that they also inhibit the cholinesterase of the nematodes.

The electrical activity in the muscle cells of *Ascaris* has been studied using microelectrodes inserted into the contractile and the non-contractile parts of the cell. The non-contractile part of the cell has a resting potential of 20 to 30 mV and this is interrupted by rhythmical spike potentials which originate at the nerve and are conducted along the innervation processes. The contractile part of the cell has a resting potential of 40 to 60 mV. Electric pulses applied to the ventral cord give rise to muscle potentials similar to the spontaneous ones and these can be recorded in muscles up to 1 cm. away from the stimulating electrodes. The wave of excitation responsible for these spikes is conducted at a velocity of about 6 cm./sec.[35, 36, 57]

Recordings made of the potentials in two muscle cells, one of which is in the dorsal part of the worm and the other in the ventral, show that there is no correlation between the timing of the two sets of pulses. Recordings from muscle cells a few millimetres apart in the same part of the nematode (i.e. dorsal or ventral) show a correlation between the two records. In muscle cells at the same distance from the head, but at different distances laterally from the longitudinal nerve, the muscle cell nearest the nerve gives the first pulse. If, on the other hand, the cells are parallel to the nerve, one being anterior to the other, then the anterior muscle gives the first pulse. A transverse cut between the two muscle cells, one anterior to the other, does not destroy the correlation between the pulses of the two muscle cells. If, however, the nerve lying parallel to the two muscle cells is cut between them, then the correlation is abolished and independent pulses are recorded from the two cells. This indicates that the longitudinal nerves in the dorsal and ventral lines co-ordinate the activity of the muscles in the dorsal and ventral parts of the nematode respectively.[68]

Ascaris possesses both inhibitory and motor nerve fibres as electrical stimulation in the region of the nerve ring (at frequencies above 5/sec.) causes inhibition of both spontaneous and electrically-stimulated contractions of the body of the nematode. Inhibition also occurs in response to shocks shorter than 0.1 m. sec. which implies that the effect is mediated by nerves. Cutting the ventral cord abolishes 80 per cent. of the inhibitory effect on body contractions of stimuli to the region of the nerve ring. Cutting the dorsal cord reduces the inhibitory effect further.[57] Inhibitory fibres can also be stimulated in pieces of *Ascaris* by electrical stimulation of the ventral nerve cord.[35, 36]

Using nerve-muscle preparations consisting of the whole larva of *Phocanema decipiens*, which encyst in the tissues of fish, observations have been made on the response of the larva to electrical stimulation and on the occurrence of spontaneous rhythmical activity when the temperature is lowered from 20° C. to 0° C.[10, 11] At 20° C. the larva of *Phocanema* contracts in response to anodal polarization (when the anode touches the worm) and relaxes in response to cathodal polarization (when the cathode touches the worm). There is a response to an anodal stimulus at all temperatures between 20° C. and 0° C. but rhythmical activity ceases as the temperature is lowered. A cathodal stimulus first causes relaxation, then no response and finally contraction as the temperature falls. The temperature at which relaxation ceases has no relation to the temperature at which rhythmical activity ceases. This suggests that relaxation and rhythmical movements are the result of two independent processes, although they may have the same effector system. Relaxation may be effected by stimulation of the muscle through the motor nerve (which may be an inhibitory nerve), and rhythmical activity by direct stimulation, possibly by stretching of the muscle cell. Thus relaxation is not an integral part of spontaneous rhythmical activity as the two can be blocked independently.[10]

Effects of drugs. Strips of body wall from *Ascaris*, consisting of muscle, nerve, hypodermis and cuticle have been used in physiological and pharmacological experiments. Many of these preparations tend to remain contracted, relaxing infrequently and only partially, which suggests a nervous upset. This is similar to the polyneuritis associated with vitamin B_1 deficiency in other animals for when small amounts of thiamine are added to the medium the strips relax more fully and more frequently and a smooth, long-lasting activity is obtained.[2]

Acetylcholine and related compounds cause contraction of the muscle of *Ascaris* when applied to isolated strips of body wall and relaxation occurs when the drug is washed away.[100] The contraction developed in these strips on addition of acetylcholine to the medium is shown in fig. 31. The extent of the contraction is directly related to the concentration of acetylcholine used. Eserine (which destroys cholinesterases) enhances the response of *Ascaris* muscle to acetylcholine, while d-tubocurarine blocks the response. The anthelmintic drug, piperazine (which lowers the response of *Ascaris* muscle to acetylcholine (fig. 31)), blocks the nerve-muscle junction or the nerves. This paralyses the nematode, so that it is swept out of the intestine of the host.[100] When piperazine is added to the bathing medium the amplitude of the spontaneous spike potentials in the muscle cells decreases until the muscle becomes electrically silent. This is accompanied by an increase in the resting potential. However, the muscle will still respond to electrical stimulation. Thus, the action of the drug is similar to that exerted by inhibitory transmitter substances on the transmembrane potential of vertebrate smooth and cardiac muscle cells. In the presence of piperazine the strength-duration curve for electrical stimuli at threshold is shifted to the right and this is

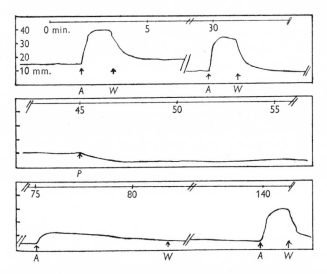

FIG. 31. Tracing from Sanborn recording of a split preparation of *Ascaris* showing the response of the preparation to acetylcholine (A) (1 μg./ml.), to washing (W), and the effect of piperazine (P) (250 μg./ml.) on the response to acetylcholine (after Norton and De Beer[100]).

consistent with a failure to respond to stimulation through the motor nerves. Piperazine does not interfere with the function of the inhibitory nerve fibres.[36, 57]

Isolated *Ascaris* muscle fails to react to adrenaline, histamine, pilocarpine and strychnine (substances which normally have a pharmacological effect on nerve-muscle preparations of other animals). Acetylcholine produces a typical stimulation as does choline, but the effects of the acetylcholine do not fall off with time,[2] probably because of the small concentration of cholinesterase

P.N.—H

present in the tissues.[17]   Nicotine has the same effect as acetylcholine.

The anterior and posterior ends of whole living *Ascaris* are very sensitive to various pharmacologically active substances. Acetylcholine (10^{-11}) applied to the lips and tail causes sharp contractions of the nematode, and after pre-treatment with pro-stigmine these parts of the nematode become sensitive to dilutions as great as 10^{-21}.   On the whole *Ascaris*, nicotine at 10^{-5} to 10^{-7} concentration stimulates activity but at higher concentrations it brings about irreversible contraction of the worm and loss of electrical excitability.   Adrenaline, muscarine and pilocarpine stimulate activity of the whole worm.   Muscle extracts and pseudocoelomic fluid taken from *Ascaris*, which has been previously treated with prostigmine for 24 hours, stimulate whole *Ascaris* and also stimulate isolated dorsal muscle preparations from the leech, showing that there has been a build up of acetyl-choline in the tissues of the prostigmine-treated *Ascaris*. Extracts from various parts of *Ascaris* inactivate 10^{-7} acetylcholine.   There is complete inactivation of the acetylcholine by pseudocoelomic fluid at 10^{-1} dilution, by extracts from the head ganglia at 10^{-3} dilution and by extracts from the lips at 10^{-5} dilution.   These results show that cholinesterase is present in *Ascaris* especially in the sensory terminals and the nerve cells.[76]

Various chemicals, which are known to have a pharmacological effect on nerve-muscle preparations of other animals, have also been tested on the larva of *Phocanema*.[11]   The effects of these chemicals on the nerve-muscle preparation are complex and are not related to their hyperpolarizing or depolarizing action.   For example, succinyl choline, nicotine and d-tubocurarine, which are all depolarizing agents, have different effects on the frequency of spontaneous rhythmical activity and on the degree of tonicity. These results do not agree with the theory that all drugs of this type act through a membrane system and that depolarization of the membrane initiates contractions while hyperpolarization brings about relaxation.[11]

Succinyl choline increases the frequency of rhythmical contraction without altering tonicity, but gamma-amino-butyric acid (GABA) and serotonin decrease tonicity, which suggests that rhythmical activity and tonic contractions are brought about by two independent systems. It is not known whether these systems are present in each muscle cell or whether the fast and slow contractions are produced by different muscle cells.[11] These results are further evidence, however, that there is an inhibitory nerve supply to the muscles of the body wall of nematodes.

When *Ascaris* is cut in half and the cut ends connected to a water manometer it lies inert at low pressures (up to 15 mm. mercury). At 45 mm. Hg occasional contractions occur, and between 60 to 100 mm. Hg regular continuous bursts of rhythmical activity occur every 20 seconds. Above 150 mm. Hg, activity becomes irregular and at 200 mm. Hg no contractions occur.[61] Similarly, body wall preparations from the anterior end of *Ascaris* show rhythmical contractions against a load of 10 to 20 gm. but not against 25 to 30 gm.[61] A tension of 10 to 20 gm. corresponds to an internal pressure of 75 to 150 mm. Hg. Contractions of the body wall muscles in one part of the nematode bring about movement of the pseudocoelomic fluid and change the internal pressure in other parts of the nematode. When the pressure in these regions reaches a certain level the muscle cells in that part of the worm are stimulated to contract directly without involving the nervous system, as suggested by the experiments on *Phocanema*. During locomotion the muscles are possibly stimulated to contract at the anterior end of the worm by the nerve ring, and their contractions stretch posterior muscles which are then stimulated to contract. In this way a wave of contraction could pass down the length of the nematode. With such mechanical co-ordination local reflex networks would be unnecessary; the simplicity of the nervous system in nematodes may be related to this.

Although little work has been done on nervous control of muscular activity in nematodes and although the muscles are morphologically unlike those of other animals, they are apparently similar pharmacologically to vertebrate skeletal muscle in that acetylcholine, and possibly serotonin, act as transmitter substances. There is also evidence for the presence of an inhibitory nerve supply to the muscles of the body wall. Rhythmical activity and tonic contraction are caused by two independent systems.

Sense Organs

Mechanoreceptors

   Labial and cephalic papillae. The mouth of nematodes is primitively surrounded by six lips, each of which is supplied with an inner and an outer labial papilla (fig. 32). Behind the lips are four cervical papillae. There are, therefore, three circles of sensory papillae on the head of the ' primitive ' nematode, namely, the inner labial papillae (6), the outer labial papillae (6) and the cervical papillae (4). This distribution of sense organs is frequently found in marine nematodes although in many

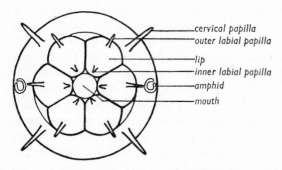

FIG. 32. *En face* view of a nematode head showing the positions of the mouth, lips amphids and papillae (Jones, *Plant Nematology*. H.M.S.O.).

terrestrial species and in nematodes parasitic in animals these sense organs are reduced in size. In some animal parasites they are also reduced in numbers, although the nerves of the ' primitive ' arrangement can still be traced (fig. 33).

   In marine nematodes the labial and cervical sense organs are often present as long bristles, with the inner circle of bristles much smaller than the outer. These bristles are cuticular structures and each is supplied by a branch from a papillary nerve. They are thought to be similar in function and physiology to the trichoid sensillae of insects (i.e. tactile structures). In animal parasitic nematodes, such as *Ascaris*, the papillae are present as small protuberances or as small pits and are innervated by branches from the papillary nerves which are known to be cholinergic. In

Ascaris the inner circle of labial papillae are lost although the nerves are still present and the three lips (formed by fusion of the original six) possess ten papillae (fig. 33). The outer labial papilla consists of a nerve fibre which narrows near the surface and ends in a sensory plate beneath the thinned cuticle. The large, outer cervical papilla consists of a deep invagination in the cuticle and a fibre projects into a pit. This fibre connects at the base of the pit

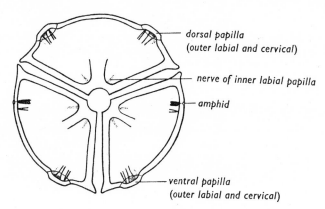

FIG. 33. *En face* view of head of *Ascaris* showing the positions of the amphids and the papillae. The dorsal cervical papillae and the dorsal outer labial papillae have fused to form two dorsal papillae, although the nerves are still separate. The ventral cervical and outer labial papillae have fused to form two ventral papillae. The inner labial papillae are lost but the nerves which originally supplied the papillae are still present. The six lips have fused to form three lips (Lee[82]).

with a branch from the papillary nerve. Pharmacologically active compounds applied to the lips of *Ascaris* cause the nematode to contract sharply, whereas there is a much weaker response when they are applied to the main body of the worm. This suggests that the papillae, or the amphids, are sensitive to these compounds.[76]

The head papillae of nematodes are sensory and probably tactile in function and thus assist them in exploring the environment and in feeding.

Campaniform-type sense organs. Sense organs, similar to the campaniform sensillae of insects, are present in certain free-living

nematodes.[67] Two types of campaniform organs have so far been
found. The first type (fig. 34*C*, *D*, *E*) consists of an elliptical cup,
into which projects a thin sheet of cuticle running longitudinally
and which is attached to the bottom and sides of the cup. This
sense organ occurs in longitudinal rows which run the whole
length of the nematode, the longitudinal axis of the sense organ
running transversely to that of the nematode. The second type
of sense organ is restricted to a short row of seven organs running

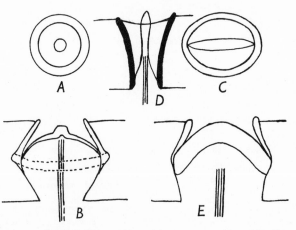

FIG. 34. Campaniform-type sense organs. *A* and *B* show
one type found only in *Chonialaimus*. *C*, *D* and *E* show a
more common type (after Inglis[67]).

posteriorly from the amphids on the head on the lateral sides of
the body. Two similar rows run posteriorly from the cloacal or
anal opening. This second type is circular in plan (fig. 34*A*, *B*)
and has a central, raised dot of thickened cuticle. A sense cell is
attached to the central part of the thickened cuticle of these sense
cells and there also appears to be a gland cell associated with them.

Campaniform sense organs are assumed to be mechanoreceptors and, because of their close resemblance to the
campaniform sensillae of insects, they probably have a similar
function (i.e. to act as proprioreceptors which record directions
of strain in the cuticle). The second type occur on the parts of the
body which are subjected to the most movement during the search
for food and during copulation.[67]

Deirids. The deirids are conspicuous papillae situated on the body of the nematode. They consist of raised and thickened areas of the cuticle supplied with a sensory nerve, and presumably are tactile structures. Nematodes also have numerous free nerve endings which appear to have a sensory function.

Genital papillae. Several papillae are found around the cloaca of males and are obviously tactile in function; they probably assist the male during copulation. The papillae consist of raised areas of thin cuticle innervated by one to three nerve fibres which end in a narrow canal. This canal opens to the exterior through a central opening in the papilla.

Chemoreceptors

The amphids and the phasmids, at least in free-living nematodes, may function as chemoreceptors but no one has verified this assumption. Some plant parasitic nematodes exhibit chemotaxis and presumably possess chemoreceptors but their location is not known. The amphids and phasmids may secrete osmotically active materials for they enlarge when certain free-living nematodes are placed in hypertonic solutions.[97]

Amphids and phasmids are paired structures found on the head and the tail respectively (fig. 1). The amphids are cuticular pits which vary in shape in different nematodes. In some free-living forms these structures are elaborate spirals, whereas in most parasitic nematodes they are simply slit-like openings. Internally the amphid consists of a pouch supplied with several nerve endings from the amphidial nerve which originates in the lateral ganglia of the nerve ring. The amphidial gland, which runs alongside the pharynx, also opens into this pouch. The amphids are well developed in free-living nematodes but are greatly reduced in animal and some plant parasitic nematodes.

The amphidial glands of *Ancylostoma caninum* contain a substance which inhibits coagulation of host blood and is thought to be secreted by the nematode during feeding.[149] Thus, the amphids of animal parasitic nematodes may have become adapted to functions other than those in free-living nematodes, assuming that in the free-living species they act as chemoreceptors.

The phasmids are paired lateral sense organs, similar to the amphids in structure, but situated posteriorly to the anus. They are absent in the Adenophorea* (Aphasmidia) which includes many free-living and some plant parasitic nematodes, but are usually present in animal parasitic species. Their structure is similar to that of the amphids, consisting of a small pouch or canal which

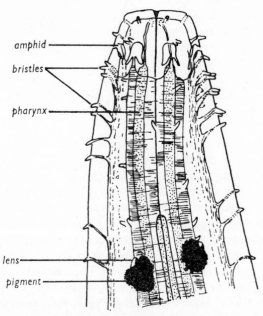

FIG. 35. Head end of *Deontostoma californicum* (dorsal view) showing the position of the ocelli. Several sensory bristles associated with the anterior end of the nematode are also shown (after Steiner and Albin[143]).

opens to the exterior and is innervated by several nerve fibres. A unicellular gland also opens into the canal or pouch. Like the amphids they are assumed to be chemoreceptors but there is no evidence for this assumption and their function is unknown.

Photoreceptors

Nematodes, like many other invertebrates, have a general sensitivity to light although photoreceptors have been found in

* Nematodes are divided into the Adenophorea or Aphasmidia and the Secernentea or Phasmidia.

only a few fresh water and marine species. A red pigment is associated with the anterior end of some free-living nematodes but the function of this pigment is not known. The adult female *Mermis*, unlike that of the young female and the male, contains a reddish pigment at the anterior end near certain cephalic nerves. This pigmentation may help the female to orientate itself when ascending the aerial parts of plants to lay eggs.[29]

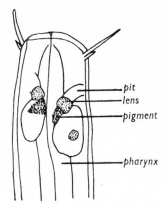

FIG. 36. The head of *Parasymplocostoma formosum* showing the position of the two ocelli (after Schulz[133]).

More highly developed, light-sensitive organs are found in certain other nematodes. In *Deontostoma* (fig. 35) and *Enoplus* the pigment of the eyespots, situated on the sides of the pharynx, is more concentrated than in *Mermis*.[143] The two ocelli of *Parasymplocostoma* are rather like those of rotifers. They are situated on the sides of the pharynx, at the bottom of a pit which opens to the exterior, and consist of a cuticular lens behind which is a layer of pigment (fig. 36).[133] The ocelli of *Leptosomatum* also have lenses but are completely enclosed in the wall of the pharynx and the lens is formed from the external covering of the pharynx.[26] Nothing is known about the innervation or the physiology of these ocelli.

7 : Locomotion and Behaviour

The Hydrostatic Skeleton

Nematodes differ from other animals in having only longitudinal muscles in the body wall. These muscles are opposed by the hydrostatic pressure of the body contents and the elasticity of the cuticle, which restore the body shape and bring the muscles back to their resting length when they relax. Because the fluid in the pseudocoelom is incompressible, and thus transmits pressure in all directions, contraction of any one muscle affects all the others, either by altering their length or the tonus which they are required to exert. In addition each muscle when it contracts opposes all the others.[25]

In *Ascaris* the success of the hydrostatic skeleton in locomotion depends on the presence of an elastic cuticle, the spiral basketwork of fibres in the fibre layer, and also on the free movement of fluid along the length of the pseudocoelom. The fibres are inextensible and allow extension and shortening of the cuticle by a lazy-tongs-like action, as they enclose between them a system of minute parallelograms (fig. 37) which can be distorted without altering the length of the sides of the parallelograms. Experiments with *Ascaris* have shown that great and often rhythmical variations in pressure occur which are related to movements of the nematode. Lengthening of the head region after shortening of the tail region is a consequence of the free transmission of hydrostatic pressure throughout the pseudocoelom.[61]

The similarity of form among nematodes may be imposed by mechanical factors because of the great internal pressure of the hydrostatic skeleton and the possession of only longitudinal muscles. The pseudocoelomic fluid in *Ascaris* is under a pressure of 70 mm. of mercury so that the intestine is normally collapsed and can be filled only by the pharynx producing higher pressures to force food into the intestine. The contents of the intestine would be forced out by the internal pressure if the pharynx and anus were not provided with ' self-sealing ' devices, which require

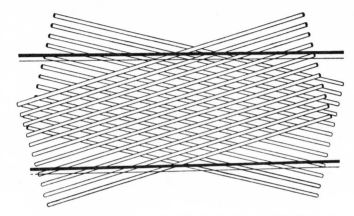

FIG. 37. A schematic representation of the spiral basketwork of fibres in the fibre layer of the cuticle of *Ascaris* illustrating the orientation of two of the fibre layers in relation to two of the transverse external annulations (Harris and Crofton[61]).

muscular action to open them. The structure of the reproductive system may also be related to the pressure in the pseudocoelom. The excretory canals, as found in *Ascaris*, are embedded in the lateral cords and are not collapsed by pressure from the pseudo-coelomic fluid. This arrangement could provide effective filtra-tion of excretory products against internal colloid osmotic pressure.[61]

Locomotion

Most nematodes move by dorso-ventral undulation. Other types of movement are uncommon and have been little studied. Stauffer[142] described annelid-like, caterpillar-like and looping progression.

Annelid-like movement

This type of movement is found in a soil inhabiting plant parasitic nematode (*Criconemoides*), which has strong, backward directed annulations which assist in propulsion (fig. 38). During

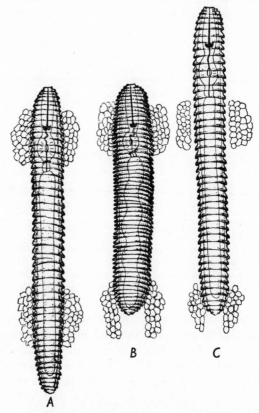

B   C

A

FIG. 38.   Locomotion of *Criconemoides* in soil.   (For description see text.)   (Stauffer[142].)

movement the whole worm shortens and thickens slightly ; the front half is prevented from reversing because the backwardly directed annulations anchor in the soil and the rear half moves forwards.   Upon relaxation the hydrostatic skeleton extends the nematode, to its resting length; however, the rear half is now anchored in the soil by the annulations and so the front half extends forwards.[142]

Recent work using ciné-film has shown that in water *Criconemoides* moves forwards by means of wave-like contractions that

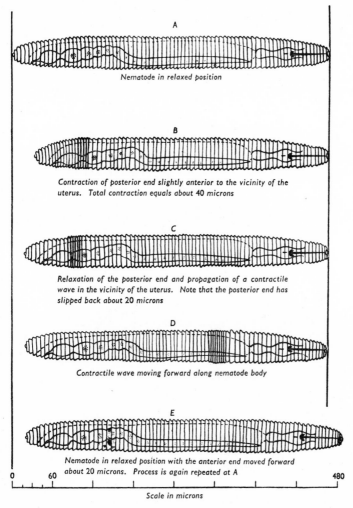

FIG. 39.   Locomotion of *Criconemoides* in water on a slide (compare with fig. 38).   (Streu, Jenkins and Hutchinson, 1961.   *New Jersey Agric. Exp. Sta. Rutgers. Bull.* 800.)

pass along the body from the posterior to the anterior end (fig. 39). Only one such wave traverses the nematode at any time. This method of progression (which is somewhat similar to the movement of *Desmoscolex* described below) moves the nematode forwards 20 μ at a time.[152] These observations, which are not in agreement

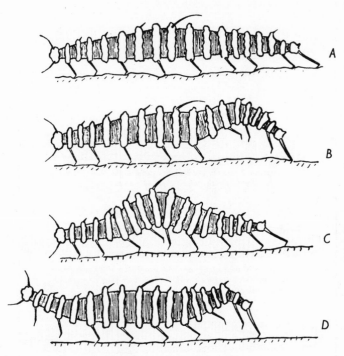

FIG. 40.   Locomotion of *Desmoscolex* (for description see text).   (Stauffer[142].)

with those of earlier workers (fig. 38), were made on glass slides and not amongst soil particles.   More work needs to be done on movement in this species, especially as Wallace [152] has found that some species of *Criconemoides* move by undulation.

Caterpillar-like movement

Desmoscolex, which has pronounced cuticular annulations with depressed rings of softer cuticle in between (fig. 40), resembles a caterpillar both in appearance and in locomotion. Waves of contraction pass from tail to head and draw together the annulations. The front is prevented from moving back by bristles which anchor in the soil. When a part of the worm contracts the rear of this part is pulled forwards and anchors in the soil, resulting in movement forwards of the worm.[142]

Looping

Chaetosoma moves like a looper caterpillar (Geometroidea). It has a few hollow bristles at the anterior end and two rows of hollow

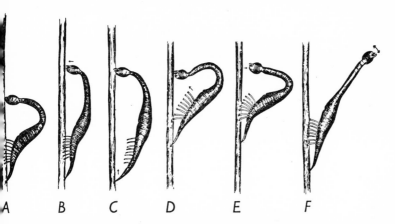

FIG. 41.   Locomotion of *Chaetosoma* on seaweed (for description see text).   (Stauffer[142].)

bristles on the ventral surface towards the rear (fig. 41). The nematode attaches itself to the substratum (seaweed) by means of these bristles, which secrete a sticky substance. During movement the ventral muscles draw the rear end towards the head (which is attached to the substratum). The bristles at the rear attach to the substratum, the front is released (possibly pulled off by contraction of the dorsal muscles) and the nematode is extended, probably by the hydrostatic skeleton and the dorsal muscles, and the process is repeated.[142]

The movement of *Criconemoides*, *Desmoscolex* and *Chaetosoma* requires a specialized arrangement of the nervous system, although co-ordination could also be achieved by pressure changes in the pseudocoelomic fluid, by mechanical stretching of adjacent muscles or by a combination of the two. Unfortunately nothing is known about the nervous systems of these nematodes.

Undulatory movement

Undulatory movement varies from a type of gliding or crawling

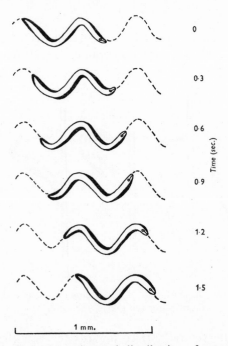

FIG. 42. Wave formation and distribution of water (black) along the body of *Aphelenchoides ritzemabosi* moving in a thin film of water. The posterior edge of each wave pushes against the water film which exerts an equal, but opposite, thrust on the nematode. The drawings were taken from a ciné film (Wallace, 1959. *Ann. appl. Biol.*).

over relatively rigid media to free-swimming, depending upon the amount of moisture present, the nature of the environment and the species of nematode. Most work has been done on the free-living stages of plant parasitic nematodes, but Wallace and Doncaster[153] showed that the locomotion of free-living species and the free-living stages of plant parasitic and animal parasitic nematodes is essentially the same.

In undulatory movement a series of waves passes posteriorly along the length of the nematode; nematodes moving between intestinal villi, through the soil, between the folded leaves of buds or within the air spaces of plants move by this means. It is essentially similar to that of snakes gliding between stones and Ceratopogonid (arthropod) larvae creeping over sand grains. In this form of locomotion each part of the animal follows the same path as the part immediately in front and the body moves along a sinuous path (fig. 42). The lateral curvature of each part of the body continuously changes to that of the part immediately in front, resulting in a bending wave being transmitted posteriorly along the length of the animal. To form bending waves the body must be subjected to appropriate bending couples. In snakes the chain of vertebrae prevent any compression of the longitudinal axis of the body and so any muscular tension developed unilaterally about this axis induces equal but opposite bending couples in the vertebrae, while in arthropods this function is performed by the exoskeleton. In nematodes and in annelids compression of the axis is resisted by the hydrostatic skeleton.[58]

The amount of moisture present in the environment has an important bearing on the locomotion of nematodes. This is because such small organisms are affected by surface tension forces, which vary according to the thickness of the water film, and as a result the nematodes are held against the substratum to a greater or lesser extent. In a particulate medium (such as soil), or in thin films of moisture on mineral, animal or plant surfaces, external resistances prevent the body of the nematode moving at right angles to its own longitudinal axis. The bending waves thus pass posteriorly along the body so as to propel the nematode forwards, the propulsive thrust necessary to maintain motion being derived from these external forces acting at right angles to the surface of the nematode. In very thin films of water forward progression almost ceases as the nematode is held firmly against the substratum by surface tension forces (fig. 43A). Greater propulsive power is obtained in such films by increasing the angle and number of waves passing along the body of the nematode and decreasing the wavelength. In slightly thicker films of water (fig. 43B) each part of the nematode follows the same path as that immediately anterior to it, so that the body moves along a sinuous track (figs. 42, 43*B*).

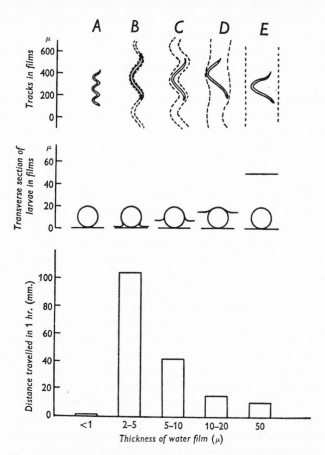

F<small>IG</small>. 43.   Movement of larvae of *Heterodera schachtii* in different thicknesses of water film on glass showing the relationship between thickness of the film and nematode mobility (Wallace, 1958. *Ann. appl. Biol.*).

The animal thus moves forward, relative to the ground, at the same speed as the wave passes back along the body.[58, 152]

Nematodes in a film of moisture of the same depth as the body diameter (fig. 43*D*) are also held in contact with the substratum, but the thicker film influences progression.   In such a film of moisture, and also in deep water (fig. 43E), the waves passing along the body also move, or slip, backwards relative to the ground.   This is because forces acting at right angles to the body can only arise when part of the body is moving with its surface inclined at an angle to the path of motion.   In this type of movement the rate of forward progression of the nematode is equal to the difference

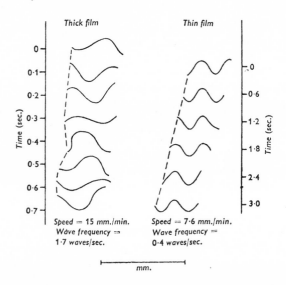

FIG. 44.   The wave motion and speed of adults of *Aphelenchoides ritzemabosi* in thick and thin films of water (Wallace, 1959.   *Ann. appl. Biol.*).

between the speed of propagation of the waves over the body and the velocity of the slip.   The maximum speed forwards must, therefore, be less than that at which the waves are travelling backwards along the body.   The motion of the wave slip relative to the long axis of the nematode is thus responsible for propelling the worm forwards.[58]

In deep water, free-living species and the free-living stages of plant and animal parasitic nematodes all exhibit similar movements. One asymmetrical wave is formed at a time, and decreases in amplitude as it passes back along the body (figs. 43E, 44). During this movement in deep water there are two points in space (nodes) where there is little lateral movement of the body of the nematode.[58, 153]

As nematodes have the same wave pattern when swimming in deep fluids, the speed of the nematode in this type of environment is a function of body length and wave frequency. The larvae of *Turbatrix aceti* and *Heterodera rostochiensis* are of about the same length (0·5 mm.) but have 250 and eight waves of contractions per minute respectively passing along their bodies. As a result *Turbatrix* travels 16 mm. while *Heterodera* travels only 0·7 mm. per minute. Similarly, nematodes with equal wave frequency but differing body length have different speeds. In deep water, nematodes with the greatest wave frequency (i.e. length × wave frequency of more than 100) are active swimmers and can swim to the surface of liquids. *Turbatrix*, which lives at the surface of malt vinegar, and *Monhystera*, which lives in ponds and streams, are included in this group. Nematodes which have a length × wave frequency between 20 and 80 are not active enough to swim upwards in deep fluids but are able to escape from the soil and ascend plants in films of moisture. Included in this group are plant parasitic nematodes which attack aerial parts of plants (e.g. *Aphelenchoides ritzemabosi*), some microbivorous species (such as *Rhabditis*, *Diplogaster* and *Plectus*) and larvae of some animal parasitic nematodes which migrate from the faecal mass, in which they have developed, and migrate to the soil surface or ascend blades of grass (e.g. *Ancylostoma*, *Necator*, *Nippostrongylus*, *Trichostrongylus*, *Haemonchus*). Maximum speed in these species is developed in thick films of moisture when contact is retained with the substratum and purchase is obtained against numerous projections (fig. 44).[153] Nematodes with a length × wave frequency less than 20 in deep water are not sufficiently active to travel far in thick water films. They move most rapidly in thin films of water and tend to live in the soil (e.g. *Heterodera* larvae) (fig. 43B).

The habitat of the free-living stages of nematodes is therefore partly related to their propulsive power which is determined by their size and activity.

During swimming, gliding and crawling the formula Vn = Vw—Vs gives the speed of a nematode, where Vn = the velocity of the nematode; Vw = the velocity of the waves along the body; Vs = slip, i.e. the velocity of the waves relative to the substratum.[153]

For a more comprehensive account of nematode locomotion see Gray[58] and Wallace.[152]

Movement in soil

The movement of nematodes in soil is partly dependent on particle size, partly upon the amount of water present in the pore spaces (fig. 45) and probably also on soil geometry and the ' mean free path ' permitted by the soil labyrinth.   Experiments using ' soils ' composed of graded sand grains showed that soil inhabiting nematodes attain maximum speeds when the channels connecting pore spaces in the ' soil ' are about the same diameter as the nematode (fig. 45).   In wider channels progression is slowed down as there is less restriction of lateral movement which reduces progression.   Channels which are narrower than the body diameter of the nematode act as barriers to movement as the nematodes seem unable to disturb the surrounding particles by burrowing. The amount of water present in the pore spaces has an important bearing on progression.   The greatest speeds occur when the pore spaces are more or less empty of water but relatively large amounts of water have collected where the particles touch (fig. 45B).[152]

Soil inhabiting nematodes move fastest when (1) there are few pore spaces narrower than the width of the nematode; (2) the pore diameter is narrow enough to restrict lateral movement; (3) the tortuosity of the channels between particles is such that the body of the nematode has waves of long wavelength and short amplitude.[152]

Movement above soil level

Nematodes are only able to move on the soil surface or on the outside of plants or decaying organic matter when there is a continuous water film or closely scattered water droplets.   These conditions are not always present, however, as the soil surface alternates between being dry, especially in the summer months, and being moist enough to allow movement to occur (e.g. after rain or dew).   Nematodes which live in this environment are (1) microbivorous feeders, (2) plant parasitic species which attack the aerial parts of plants and (3) larvae of animal parasitic nematodes which have reached the infective stage after a period of growth and

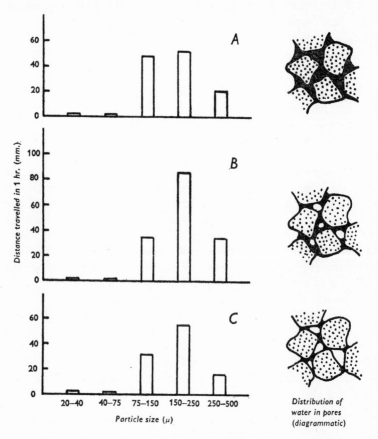

FIG. 45. Migration of larvae of *Heterodera* in different fractions of soil at three different moisture levels. The diagrams to the right of the histograms indicate the distribution of water in pores at the three moisture levels (Wallace, 1958. *Ann. appl. Biol.*).

development in the soil or the faecal mass and are migrating on to the soil surface or ascending herbage.

Aphelenchoides ritzemabosi moves fastest on the chrysanthemum when it is moving between epidermal hairs in a thick film of water. Movement is slower on surfaces with few epidermal hairs or in thin films of moisture. In thick films of moisture, forward progression is not uniform (dotted line in fig. 44) but the nematode is very active and speeds of 15 mm./min. are possible. In thin films, however, uniform motion occurs (fig. 44) but the speed at which the nematode travels forwards is slower than in a thick film. This species is thus better adapted to swimming (i.e. progression by two-way undulatory propulsion with wave slip) than to gliding.[152]

The free-living infective larvae of animal parasitic nematodes spend the first part of their life in the faeces or in the surrounding soil and are essentially soil dwelling types. The infective larva, however, is a migratory stage and usually moves out of the faeces. The larvae of most species move randomly in a continuous water film but in narrow channels of moisture the larvae tend to continue movement in the direction of the channel. The narrower the channel the further the larva travels along it because the possibility of random movement decreases. Thus, on a flat horizontal surface such as soil, the larvae move until they make contact with a vertical projection, such as a blade of grass.[31] Larvae of *Trichostrongylus* are apparently attracted to the roots of grasses.[153] The migration of the larvae on the blade of grass depends on the width of the vertical path; a narrow blade tends to guide the larvae upwards or downwards according to its orientation. Larvae tend to be trapped at the tip of the grass if there is little moisture. A continuous film of moisture on broad, smooth blades of grass reduces vertical movement, whereas narrower, or veined leaves in which moisture collects between the veins, allows greater vertical movement.[31] Larvae that migrate up the aerial parts of plants must swim efficiently in thick films of moisture and it has been shown that the migratory larvae of a sheep nematode (*Trichostrongylus colubriformis*) and of a rat nematode (*Nippostrongylus brasiliensis*) have a length × wave frequency between 20 and 80, which places them in the group of active swimmers in thick films of moisture.[153] The tendency to ascend herbage is of survival value to those nematodes which have to be eaten by a grass eating host for completion of the life cycle to occur.

Some nematodes ascend plant stems (*Mermis* females) or the

hairs of the animal host (*Nippostrongylus* larvae) by means of three-dimensional undulatory propulsion.[58]

Movement inside animals

Little is known about the movement of nematodes inside other animals. X-ray studies of *Ascaris* in man show that the nematode remains stationary most of the time, being braced against the walls

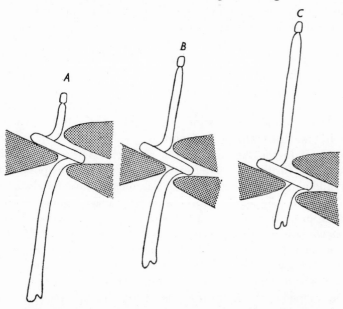

Fig. 46. The locomotion of an adult *Nippostrongylus* removed from the intestine of the host and placed amongst moist sand grains. It is probable that similar movements are performed by the nematode amongst the villi of the host intestine (Lee[83]).

of the intestine and not affected by normal movements of the organ. The rather ineffectual movements of *Ascaris* lying free in saline, contrast with the spiral forward movement in the intestine. This suggests that locomotion is, at least partially, by three-dimensional undulatory propulsion.[88] Adult *Nippostrongylus*, when moving in a viscous medium or amongst solid particles, moves by two- or three-dimensional undulatory propulsion. In three-dimensional propulsion a single spiral, corkscrew-like wave passes backwards along the length of the worm. This spiral wave obtains purchase against solid particles in the environment and the position of the wave remains constant to the substratum (fig. 46).[83] This type of movement probably occurs in the intestine of the host.

Passive movement

The passive dispersal of nematodes is an important means of ensuring the spread of the species. This is especially true for soil inhabiting and plant parasitic nematodes which travel at the most only a few yards each year by their own movements. It is not so important in nematodes which parasitize animals because the movements of the host from place to place ensure the dispersal of the parasite.

The means of disseminating nematodes passively are numerous. Whilst some methods are localized in their effects others are responsible for the spread of nematodes over wide areas. Most passive dispersal is by the lateral movements of eggs and cysts brought about by natural and artificial agencies. Thus eggs of *Heterodera schachtii*, contained in cysts, are spread by general farm activities (when infected soil adheres to the boots of farm workers, to the feet of animals and to farm implements); by infected soil adhering to seed, tubers, bulbs and all kinds of transplants; by wind (when soil containing eggs and cysts of nematodes is carried away in dust storms); by water currents and the run off of surface water; and by the activities of man in transporting soil containing cysts for great distances. Many plant parasitic nematodes (e.g. *Radolphus similis* on banana ' sets ', *Heterodera rostochiensis* cysts on potato tubers) are spread with their hosts, so that they both tend to have the same distribution except where some climatic or other factors operates against the nematode. In contrast to mobile animals such as airborne aphids, which colonize great tracts of country and may sometimes affect almost every available host plant therein within a matter of weeks, the spread of nematodes is slow. It takes *Ditylenchus dipsaci*, for example, about three years to colonize a whole field of lucerne from restricted foci introduced on seed. This dispersal is assisted by rain splash, surface drainage, cultivation of the soil and hay making. It has taken *Heterodera schachtii* about 40 years to establish itself

in the peat fens (which earlier were swamps) of East Anglia, and
H. rostochiensis has taken 70–100 years to establish itself in
Europe from the Andes—the source of both the potato and this
associated nematode. The spread and establishment of both of
these species of *Heterodera* have been greatly assisted by man.[69, 152]

Movement of nematodes in the soil and on the soil surface is
affected by rainfall and by evaporation from the soil. During and
after rain there is a downward movement of water in the soil and
this may carry nematodes with it. The speed at which the
nematode is carried down varies with the type of soil and the
length of the nematode. Dead or inactive nematodes are not
carried through the soil as far as active ones, which suggests that
slight flexing movements of the nematode body are essential for
passage.[152]

The infective larvae of a cattle nematode (*Dictyocaulus viviparus*) are sluggish, generally inactive and have no sustained powers
of migration, yet they become widely dispersed on to the herbage
surrounding the faeces in which they develop. Their dispersal
is brought about by the fungus *Pilobolus*. The larvae develop to
the migratory infective stage in the faeces and migrate to the
surface of the faecal mass. At about the same time there is usually
abundant development, on the surface of the faeces, of the
sporangiophores of *Pilobolus*. This fungus is stimulated to dis-
charge its sporangia violently by changes in illumination and carries
larvae of the nematode, which have ascended the sporangiophores,
on to the surrounding herbage. Transport by this means is
important in the life cycle of the nematode as cattle do not feed
on grass in the vicinity of faeces.[114]

The larvae of many dung-inhabiting nematodes are transported
from old to new faeces by insects. Some become attached to
bodies or to hairs on the legs of flies, while others move under the
elytra of dung beetles and remain there until the beetle moves to
freshly deposited dung.[9]

General Behaviour

Response to gravity

Free-living stages of many animal parasitic nematodes are
apparently negatively geotactic, but upward movement of the

larvae is the result of random movements and not a response to gravity. These larvae move in all directions, even on a vertical plane, and as discussed earlier (see movement above soil level) the moisture on the grass in their habitat and the nature of the leaf surface are important factors which determine upward or downward movement. In the field, therefore, the number of larvae moving upwards is directly proportional to the number of blades of grass in the vicinity and the amount of moisture present on the grass. A plant parasitic nematode (*Ditylenchus dipsaci*) which attacks the aerial parts of the host plant, is unaffected by gravity. The surface tension forces acting on a nematode covered by a water film are 10^4 to 10^5 times greater than gravity, therefore gravity is unlikely to affect orientation of nematodes in such situations.[31, 151, 152]

Nematodes living in deep water or other liquids are subjected to appreciable gravitational forces. *Turbatrix aceti*, which lives at the surface of malt vinegar, may exhibit negative geotaxis. However, the tendency of this species to keep in the surface layers of the vinegar may be due to the fact that it is an active swimmer with a relatively heavy tail. Similarly, larvae of *Heterodera* are ' head-heavy ' and thus have the opposite tendency and do not rise through the soil moisture to the surface of the soil.[151]

Response to water currents

There is little evidence that water movements affect the orientation of nematodes (rheotaxis) other than mechanically. *Aphelenchoides ritzemabosi*, a parasite of the aerial parts of chrysanthemums, moves upwards on the stem of the plant in a stationary film of moisture. Downward currents, caused by rain or heavy dew, do not stimulate an upward migration but merely wash the nematodes down the stem.[151]

Response to an electrical field

Many free-living and plant parasitic nematodes migrate to the cathode (*Panagrellus*, *Turbatrix*, *Rhabditis*, *Dorylaimus*, *Tylenchus*, *Ditylenchus*, *Aphelenchoides*) or to the anode (*Heterodera schachtii* larvae).[23, 70] This galvanotactic behaviour is dependent on potential gradient rather than the size of the current (the minimum threshhold to which *Ditylenchus* and *Turbatrix* respond is 30 mV/mm.). Electrical potential gradients occur around the roots of plants and attraction of *Meloidogyne* larvae to plant roots may be to low oxidation-reduction (redox) potentials.[6] Plant parasitic nematodes may thus be attracted to the roots of plants along a potential gradient but only over distances of a few millimetres. They may also be attracted to certain parts of the root by the positive surface charge of the root tip and the zone of differentiation or the negative surface charge of the zone of elongation. Cell sap, which is negatively charged, when released from punctured or injured cells may also attract nematodes by this means.[70, 152]

The free-living infective larvae of at least one animal parasitic nematode (*Trichostrongylus retortaeformis*) migrate to the cathode.[60] The larvae of *Phocanema* (which are encysted in the muscles of fish and which infect seals), on the other hand, show no galvanotactic behaviour.[126]

Response to chemicals

Plant parasitic nematodes are attracted to the roots of host plants by various stimuli. The roots of plants secrete substances, which can pass through a permeable membrane, and attract nematodes to the roots. Larvae of *Heterodera* are attracted from a distance of 2 cm. to roots of the host plant and also move to a point from which a plant has been removed, showing that the attractant substance is chemical in nature.[151] Millet exudes a chemical which attracts *Hemicycliophora paradoxa* from a distance of 40 cm.[152] *Meloidogyne* larvae are strongly attracted to points on the roots where cells have been injured, indicating that cell sap is a strong attractant. This could be a chemical attraction or a galvanotactic reaction, as the sap is negatively charged.[70] *Meloidogyne* larvae are also attracted to certain reducing substances, probably because of the changes in redox potential, and to certain acidic substances, although there is no response to a pH gradient.[6]

Chemotaxis is probably responsible for the attraction of opposite sexes in mating but there is no evidence for this.

Response to gases

Different results have been recorded with different species of nematodes exposed to oxygen or to carbon dioxide gradients. Whereas many soil inhabiting nematodes are attracted to a source of carbon dioxide, such as plant roots, some (*Meloidogyne* larvae) are not. Most soil inhabiting nematodes are probably aerobic but they apparently do not respond to an oxygen gradient; *Meloidogyne* larvae move into areas of low oxygen tension in response to low redox potentials.[6, 151]

Response to a moisture gradient

Active nematodes in the soil are always in water and cannot be hydrotactic. The ability of some soil nematodes to migrate to the wet end of a moisture gradient is thought to be a mechanical response. *Heterodera rostochiensis* larvae are attracted to host roots against a water gradient.[151]

Response to touch

The skin-penetrating larvae of animal parasitic nematodes (*Ancylostoma, Necator*) respond to contact with an obstacle by orientating themselves at right angles to the surface of the obstacle and moving vigorously. This behaviour is of obvious value to larvae which penetrate the host through the skin. Similar behaviour may be exhibited by larvae of *Heterodera* and other plant parasitic nematodes when they make contact with host roots.

Thigmo-kinesis has been defined as the behaviour which determines where the animal shall stop rather than guiding an animal, as in thigmo-taxis. Nematodes which live in amongst the villi of the intestine of the animal host may be thigmo-kinetic, which would keep them close to the mucosa and stop them from wandering into the lumen of the intestine, but there is no experimental evidence for this. Marine annelids, which normally hide in crevices and under rocks, aggregate together when placed in a dish. Many nematodes show a similar aggregation when placed in a dish of water, or adhere together on the top of faeces or soil, and this may be a thigmo-kinetic response.

Response to light

A species of *Rhabditis*, which lives in mushroom beds, is positively phototactic. The nematodes move to the surface and congregate in clumps when exposed to light but disappear from the surface in the dark. The first stage larvae of *Dictyocaulus* are positively phototactic but the second and third stages are not.[151] Trichostrongyle larvae show the greatest response to dim light of 62 foot candles but respond less to stronger or weaker light.[116] Many skin penetrating larvae are supposed to be positively phototactic but the response is probably to heat from the light source. Light has no effect on the movements of the plant parasitic nematode *Ditylenchus dipsaci*.[151, 152] Several free-living nematodes possess ocelli but little work has been done on the behaviour of these species when stimulated by light. They may be positively phototactic and orientate themselves so that the ocelli on each side of the body receive equal illumination during movement. The adult female of *Mermis* may use the reddish pigment at the anterior end to orientate itself when ascending the aerial parts of plants upon which it lays its eggs.[29]

Appendix

A list of nematodes which are mentioned in this book with some short notes on their biology.

NEMATODE	BIOLOGY
Acrobeles species	Free-living in soil; microbivorous.
Actinolaimus spp.	Free-living; predatory.
Agamermis spp.	Adult free-living; larva parasitic in the haemocoele of insects.
Ancylostoma caninum	Adult parasitic in intestine of dog (hookworm); 1st three larval stages free-living; 3rd larva penetrates skin of host.
A. duodenale	Similar to *A. caninum* but parasitic in intestine of man (hookworm).
Anguina tritici	Parasitic on plants (cause of ear cockle in wheat).
Anisakis spp.	Parasitic in stomach of marine mammals and birds.
Aphelenchoides spp.	Some species are plant parasites; some feed on fungi; some are parasitic in insects.
Aphelenchus spp.	Phytophagus (probably feed on fungi).
Ascaridia galli	Parasitic in intestine of fowl.
Ascaris lumbricoides	Parasitic in intestine of pig and man.
Aspiculuris spp.	Parasitic in colon of rodents.
Angusticaecum spp.	Parasitic in alimentary canal of tortoise.
Caenorhabditis spp.	Free-living in soil; microbivorous.
Camallanus spp.	Parasitic in alimentary tract of fish, amphibia and reptiles.
Capillaria hepatica	Parasitic in liver of mammals.
Cephalobus spp.	Free-living in soil; microbivorous.
Chabertia spp.	Parasitic in alimentary tract of ungulates.
Chaetosoma spp.	Free-living; marine.
Chonialaimus spp.	Free-living; marine.
Contracaecum spp.	Parasitic in intestine of fish, amphibia and reptiles; larva in body cavity of fish.
Cooperia spp.	Parasitic in intestine of ungulates.
Cosmocerca spp.	Parasitic in intestine of amphibia and reptiles.
Criconemoides spp.	Ectoparasitic on plant roots (ring nematodes).

Deontostoma spp.   Free-living; marine.
Desmoscolex spp.   Free-living in fresh water.
Dictyocaulus viviparus   Parasitic in lungs of cattle.
Dioctophyme renale   Parasitic in kidney of mammals.
Diplogaster spp.   Some free-living (microbivorous or pre-dators); others parasitic in insects.
Ditylenchus destructor   Plant parasite (potato rot nematode).
D. dipsaci   Plant parasite (bulb and stem nematode).
D. triformis   Feeds on fungi (may be parasitic on plants).
Dirofilaria spp.   Parasitic in heart and connective tissue of mammals; pre-larvae (microfilariae) trans-mitted by mosquito.
Dolichodorus spp.   Ectoparasitic on plant roots (awl nematode).
Dorylaimus spp.   Predatory and phytophagous species occur in soil and fresh water.
Dracunculus spp.   Parasitic in connective tissue of mammals (Guinea worm); larva in copepods.
Enoplus spp.   Free-living; marine.
Eustrongyloides spp.   Parasitic in intestine of birds; larva in muscles of fish.
Graphidium strigosum   Parasitic in alimentary tract of rabbit.
Haemonchus contortus   Parasitic in abomasum of sheep and cattle (stomach worm); 1st three larval stages free-living.
Hammerschmidtiella spp.   Parasitic in hind-gut of cockroaches.
Hemicycliophora spp.   Ectoparasitic on plant roots (sheath nema-todes).
Heterakis gallinae   Parasitic in caecum of fowl.
Heterodera göttingiana   Endoparasitic on peas (pea cyst nematode).
H. rostochiensis   Endoparasitic on potatoes and tomatoes (potato root eelworm, or golden nematode).
H. schachtii   Endoparasitic on sugar beet and other hosts (sugar beet eelworm).
Leidynema spp.   Parasitic in hind-gut of cockroaches.
Leptosomatum spp.   Free-living; marine.
Litomosoides carinii   Parasitic in body cavity of rodents; trans-mitted by mites.
Meloidogyne spp.   Endoparasitic in plant roots (root-knot nematodes).
Mermis spp.   Adult free-living; larva parasitic in haemo-coele of insects.
Metastrongylus spp.   Parasitic in bronchi of mammals.
Monhystera spp.   Free-living; phytophagous.
Mononchus spp.   Free-living; predator; cosmopolitan.
Necator americanus   Parasitic in intestine of man (hookworm); life cycle similar to that of *Ancylostoma*.

Nematodirus spp.	Parasitic in intestine of ruminants.
Neoaplectana spp.	Parasitic in insects and feeds on their decaying tissues after killing insect.
Neotylenchus spp.	Free-living and endoparasitic on plants.
Nippostrongylus brasiliensis	Parasitic in intestine of rats; life cycle similar to that of *Ancylostoma*.
Nygolaimus spp.	Free-living; predator.
Oesophagostomum spp.	Parasitic in colon of mammals.
Onchocerca spp.	Parasitic in connective tissue of mammals; pre-larvae (microfilariae) transmitted by *Simulium* spp. (Black Fly).
Ostertagia spp.	Parasitic in abomasum of ruminants.
Oswaldocruzia spp.	Parasitic in alimentary tract of amphibia and reptiles.
Oxyuris equi	Parasitic in large intestine of horses.
Panagrellus silusiae	Free-living; microbivorous.
Panagrolaimus rigidus	Free-living; cosmopolitan.
Paraphelenchus spp.	Free-living in soil; fungal feeders.
Parascaris equorum	Parasitic in small intestine of horses.
Parasitylenchulus diplogenus	Parasitic in fruit flies (*Drosophila*).
Parasymplocostoma formosum	Free-living; marine.
Pelodera spp.	Free-living; microbivorous.
Phocanema decipens	Parasitic in stomach of seals; larvae in muscles of fish.
Plectus spp.	Free-living in soil, fresh water and mosses; microbivorous.
Porracaecum decipens	Parasitic in intestine of birds; larvae in earthworm.
Pratylenchus spp.	Endoparasitic in plant roots (root lesion nematodes).
Radopholus similis	Endoparasitic in plant roots (burrowing nematode).
Rhabdias spp.	Parasitic in lungs of amphibia and reptiles.
Rhabditis coarctata	Free-living in dung; larvae transported by insects.
R. dubia	Free-living in dung; larvae transported by insects.
R. terrestris	Larva parasitic in earthworms; adults and larval stages live on decaying earthworm.
Rotylenchulus spp.	Parasitic on plant roots.
Rotylenchus spp.	Parasitic on plant roots.
Spironoura spp.	Parasitic in intestine of fish, amphibia and reptiles.

P.N.—K

Strongyloides spp.	Parasitic in intestine of vertebrates; larvae free-living.
Strongylus equi	Parasitic in large intestine of horses.
Syngamus spp.	Parasitic in bronchi of birds.
Syphacia spp.	Parasitic in alimentary tract of rodents.
Thelastoma spp.	Parasitic in hind-gut of cockroaches.
Toxocara spp.	Parasitic in intestine of carnivores.
Trichinella spiralis	Parasitic in intestine of mammals; larvae encyst in muscles of mammals.
Trichostrongylus axei	Parasitic in abomasum of sheep and cattle; 1st three larval stages free-living.
T. colubriformis	Parasitic in small intestine of sheep and cattle; 1st three larval stages free-living.
T. retortaeformis	Parasitic in small intestine of rabbit; 1s three larval stages free-living.
Trichuris muris	Parasitic in intestine of rats and mice.
T. ovis	Parasitic in caecum of ruminants.
T. vulpis	Parasitic in intestine of dog.
Turbatrix aceti	Free-living in malt vinegar; microbivorous (vinegar eelworm).
Tylenchorhynchus spp.	Free-living or plant parasites.
Tylenchulus spp.	Parasitic on plant roots.
Tylenchus spp.	Free-living or plant parasites; feed on root hairs or fungi.
Uncinaria stenocephala	Parasitic in intestine of dog (hookworm); 3rd larva penetrates the skin.
Wuchereria bancrofti	Parasitic in connective tissue of man (filarial worm); pre-larvae (microfilariae) in mosquitoes.
Xiphinema spp.	Parasitic on plant roots (dagger nematode).

References

* *Denotes a paper, review or book with a comprehensive
bibliography in which references not given below can be found*

1. ACKERT, J. F. 1931. The morphology and life history of the fowl
   nematode *Ascaridia lineata* (Schneider). *Parasitology,* **23**:
   360–79.
2. BALDWIN, E. and MOYLE, V. 1947. An isolated nerve-muscle
   preparation from *Ascaris lumbricoides. J. exp. Biol.* **23**:
   277–91.
3. BECKETT, E. B. and BOOTHROYD, B. 1961. Some observations on
   the fine structure of the mature larva of the nematode *Trichinella spiralis. Ann. Trop. Med. Parasit.* **55**: 116–24.
4. BEHERENZ, K. W. 1956. Vergleichende physiologische Untersuchungen über die Exkretion parasitischer Nematoden mit
   Hilfe der Fluoreszenzmikroskopie. *Z. wiss. Zool.* **159**:
   129–64.
5. BIRD, A. F. 1959. Development of the root-knot nematodes
   Meloidogyne javanica (Treub) and *Meloidogyne hapla* Chitwood
   in the tomato. *Nematologica,* **4**: 31–42.
6. —— 1959. The attractiveness of roots to the plant parasitic nematodes *Meloidogyne javanica* and *M. hapla. Nematologica,* **4**:
   322–35.
7. BIRD, A. F. and DEUTSCH, K. 1957. The structure of the
   cuticle of *Ascaris lumbricoides* var. *suis. Parasitology,* **47**:
   319–28.
8. BLAKE, C. D. 1961. Importance of osmotic potential as a com-
   ponent of the total potential of the soil water on the movement
   of nematodes. *Nature, Lond.* **192**: 144–5.
9. BOVIEN, P. 1937. Some types of association between nematodes
   and insects. *Vidensk. Medd. dansk naturh. Foren. Kbh.* **101**:
   114 pp.
10. BRADLEY, C. 1961. The effect of electrical stimulation at low
    temperatures on the larvae of *Phocanema decipens. Canad. J.
    Zool.* **39**: 35–42.
11. —— 1961. The effect of certain chemicals on the response to
    electrical stimulation and the spontaneous rhythmical activity
    of larvae of *Phocanema decipens. Canad. J. Zool.* **39**:
    129–36.

*12. VON BRAND, T. 1952. *Chemical physiology of endoparasitic animals.* Academic Press Inc., N.Y. 339 pp.

*13. 1960. Chapters 26–29. Physiology and biochemistry of nematodes. In: *Nematology. Fundamentals and recent advances with emphasis on plant parasitic and soil forms.* Editors: Sasser, J. N. and Jenkins, W. R. The Univ. of North Carolina Press, Chapel Hill.

14. VON BRAND, T., BOWMAN, I. B. R., WEINSTEIN, P. P. and SAWYER, T. K. 1963. Observations on the metabolism of *Dirofilaria uniformis. Exp. Parasit.* **13**: 128–33.

15. VON BRAND, T., WEINSTEIN, P. P., MEHLMAN, B. and WEINBACH, E. C. 1952. Observations on the metabolism of bacteria-free larvae of *Trichinella spiralis. Exp. Parasit.* **1**: 245–55.

16. BUEDING, E. 1949. Studies on the metabolism of the filarial worm, *Litomosoides carinii. J. exp. Med.* **89**: 107–30.

17. 1952. Acetylcholinesterase activity of *Schistosoma mansoni. Brit. J. Pharmacol.* **7**: 563–6.

*18. 1962. Comparative aspects of carbohydrate metabolism. *Fed. Proc.* **21**: 1039–46.

19. BUEDING, E. and CHARMS, B. 1952. Cytochrome c, cytochrome oxidase, and succinoxidase activities of helminths. *J. biol. Chem.* **196**: 615–27.

20. BUEDING, E., KMETEC, E., SWARTZWELDER, S. A. and SAZ, H. J. 1961. Biochemical effects of dithiazine on the canine whip-worm, *Trichuris vulpis. Biochem. Pharmacol.* **5**: 311–22.

21. BUEDING, E. and OLIVER, GONZALEZ, J. 1950. Aerobic and anaerobic production of lactic acid by the filarial worm, *Dracunculus insignis. Brit. J. Pharmacol.* **5**: 62–4.

22. CARPENTER, M. F. P. 1952. The digestive enzymes of *Ascaris lumbricoides* var. *suis*; their properties and distribution in the alimentary canal. *Dissertation. Univ. Michigan. Univ. Microfilms, Publ. no.* 3729. *Ann Arbor, Mich.* 183 pp.

23. CAVENESS, F. H. and PANZER, J. D. 1960. Nemic galvanotaxis. *Proc. helm. Soc. Wash.* **27**: 73–4.

24. CAVIER, R. and SAVEL, J. 1954. L'uréogénèse chez l'*Ascaris* du porc (*Ascaris lumbricoides* Linné, 1758). *Bull. Soc. Chim. biol., Paris,* **36**: 1425–31.

*25. CHAPMAN, G. 1958. The hydrostatic skeleton in the invertebrates. *Biol. Rev.* **33**: 338–71.

*26. CHITWOOD, B. G. and CHITWOOD, M. B. 1950. *An introduction to nematology.* Monumental Printing Co., Baltimore.

27. CHITWOOD, M. B. 1951. Notes on the physiology of *Meloidogyne javanica* (Treub, 1885). *J. Parasit.* **37**: 96–8.

28. COBB, N. A. 1915. Nematodes and their relationships. *U.S. Dept. of Agriculture Yearbook for* 1914. 457–90.

29. COBB, N. A. 1929. The chromatropism of *Mermis subnigrescens*, a nemic parasite of grasshoppers. *J. Wash. Acad. Sci.* **19**: 159–66.

30. COSTELLO, L. C. and GROLLMAN, S. 1959. Studies on the reactions of the Krebs cycle in *Strongyloides papillosus* infective larvae. *Exp. Parasit.* **8**: 83–9.

31. CROFTON, H. D. 1954. The vertical migration of infective larvae of Strongyloid nematodes. *J. Helminth.* **28**: 35–52.

32. CROWLEY, K. and WARREN, L. G. 1963. Production of volatile acids by *Ancylostoma caninum*. *J. Parasit.* **49** (Suppl.)**:** 52.

33. DAVENPORT, H. E. 1949. The haemoglobins of *Ascaris lumbricoides*. *Proc. roy. Soc.* B, **136**: 255–70.

34. —— 1949. The haemoglobins of *Nippostrongylus muris* (Yokagawa) and *Strongylus* spp. *Proc. roy. Soc.* B, **136**: 271–80.

35. DEBELL, J. T., DEL CASTILLO, J. and SANCHEZ, V. 1963. Electrophysiology of the somatic muscle cells of *Ascaris lumbricoides*. *J. cell. comp. Physiol.* **62**: 159–77.

36. DEL CASTILLO, J., MORALES, T. A. and SANCHEZ, V. 1963. Action of piperazine on the neuromuscular system of *Ascaris lumbricoides*. *Nature, Lond.* **200**: 706–7.

37. DE LEY, J. and VERCRUYSE, R. 1955. Glucose-6-phosphate and gluconate-6-phosphate dehydrogenase in worms. *Biochim. Biophys. Acta*, **16**: 615–16.

38. DONCASTER, C. C. 1962. Nematode feeding mechanisms. 1. Observations on *Rhabditis* and *Pelodera*. *Nematologica*, **8**: 313–20.

39. —— (Personal communication).

*40. DROPKIN, V. H. 1955. The relations between nematodes and plants. *Exp. Parasit.* **4**: 282–322.

41. ELLENBY, C. 1946. Nature of the cyst wall of the potato-root eelworm *Heterodera rostochiensis*, Wollenweber, and its permeability to water. *Nature, Lond.* **157**: 302.

42. ELLISON, T., THOMSON, W. A. B. and STRONG, F. M. 1960. Volatile fatty acids from axenic *Ascaris lumbricoides*. *Arch. Biochem. Biophys.* **91**: 247–54.

43. ELLS, H. A. and READ, C. P. 1961. Physiology of the vinegar eel, *Turbatrix aceti* (Nematoda). I. Observations on respiratory metabolism. *Biol. Bull.* **120**: 326–36.

44. ENIGK, K. 1938. Ein Beitrage zur Physiologie und zum Wirt-Parasit-Verhältnis von *Graphidium strigosum* (Trichostrongylidae, Nematoda). *Z. Parasitenk.* **10**: 386–414.

45. ENIGK, K. and GRITTNER, I. 1952. Zur Morphologie von *Strongylus vulgaris* (Nematodes). *Z. Parasitenk.* **15**: 267–82.

46. ENTNER, N. and GONZALEZ, C. 1959. Fate of glucose in *Ascaris lumbricoides*. *Exp. Parasit.* **8**: 471–9.

*47. FAIRBAIRN, D.   1957.   The biochemistry of *Ascaris*.   *Exp. Parasit.* **6**: 491–554.

*48.   1960.   The physiology and biochemistry of nematodes. Chapter 30.   In: *Nematology. Fundamentals and recent advances with emphasis on plant parasitic and soil forms.* Editors: Sasser, J. N. and Jenkins, W. R.   The University of North Carolina Press, Chapel Hill.   480 pp.

*49.   1960.   Physiologic aspects of egg hatching and larval exsheathment in nematodes.   In: *Host influence on parasite physiology.* Editor: Stauber, L. A.   Rutgers Univ. Press, New Brunswick. 96 pp.

50. FENWICK, D. W.   1939.   Studies on the saline requirements of the larvae of *Ascaris suum*.   *J. Helminth.* **17**: 211–28.

51. FERNANDO, M. A.   1963.   Metabolism of hookworms.   I. Observations on the oxidative metabolism of free living third stage larvae of *Necator americanus*.   *Exp. Parasit.* **13**: 90–7.

52. FLURY, F.   1912.   Zur chemie and Toxikologie der Ascariden. *Arch. exp. Pharmak.* **67**: 275–392.

53. GOFFART, H. and HEILING, A.   1962.   Beobachtungen über die Enzymatische wirkung von Speicheldrüsensekretion Pflanzenparasitärer Nematoden.   *Nematologica*, **7**: 173–6.

54. GOLDBERG, E.   1957.   Studies on the intermediary metabolism of *Trichinella spiralis*.   *Exp. Parasit.* **6**: 367–82.

55.   1958.   The glycolytic pathway in *Trichinella spiralis* larvae.   *J. Parasit.* **44**: 363–70.

56. GOODEY, J. B.   1959.   The excretory system of *Paraphelenchus* and the identity of the hemizonid.   *Nematologica*, **4**: 157–9.

57. GOODWIN, L. G. and VAUGHAN WILLIAMS, E. M.   1963.   Inhibition and neuromuscular paralysis in *Ascaris lumbricoides*.   *J. Physiol.* **168**: 857–71.

*58. GRAY, J.   1953.   Undulatory propulsion.   *Quart. J. micr. Sci.* **94**: 551–78.

59. VAN GREMBERGEN, G.   1954.   Haemoglobin in *Heterakis gallinae*. *Nature, Lond.* **174**: 35.

60. GUPTA, S. P.   1963.   Mode of infection and biology of infective larvae of *Molineus barbatus*, Chandler 1942.   *Exp. Parasit.* **13**: 252–5.

61. HARRIS, J. E. and CROFTON, H. D.   1957.   Structure and function in the nematodes: internal pressure and cuticular structure in *Ascaris*.   *J. exp. Biol.* **34**: 116–30.

62. HINZ, E.   1963.   Elektronenmikroskopische Untersuchungen an *Parascaris equorum*.   *Protoplasma*, **56**: 202–41.

*63. HOBSON, A. D.   1948.   The physiology and cultivation in artificial media of nematodes parasitic in the alimentary tract of animals.   *Parasitology*, **38**: 183–227.

64. HOBSON, A. D., STEPHENSON, W. and BEADLE, L. C. 1952. Studies on the physiology of *Ascaris lumbricoides*. I. The relation of the total osmotic pressure, conductivity and chloride content of the body fluid to that of the external environment. *J. exp. Biol.* **29**: 1–21.

65. HSÜ, H. F. 1933. Study on the oesophageal glands of parasitic nematoda superfamily Ascaroidea. *Chinese Med. J.* **47**: 1247–88.

66. HURLAUX, R. 1948. Les cellules oxydasiques de l'*Ascaris*. *La Feuille des Naturalistes*, **3**: 5–15.

67. INGLIS, W. G. 1963. ' Campaniform-type ' organs in nematodes. *Nature, Lond.* **197**: 618.

68. JARMAN, M. 1959. Electrical activity in the muscle cells of *Ascaris lumbricoides*. *Nature, Lond.* **184**: 1244.

*69. JONES, F. G. W. 1959. Ecological relationships of nematodes. In: *Plant Pathology. Problems and progress 1908–1958.* Editors: Hollon, C. S., Fischer, G. W., Fulton, R. W., Hart, H. and McCallan, S. E. A. Univ. of Wisconsin Press, Madison.

70. ———— 1960. Some observations and reflections on host finding by plant nematodes. *Meded. LandbHoogesch. Gent*, **25**: 1009–24.

71. KATZ, B. 1962. The transmission of impulses from nerve to muscle, and the subcellular unit of synaptic action. *Proc. roy. Soc.* **B**, **155**: 455–77.

72. KESSEL, R. G., PRESTAGE, J. J., SEKHON, S. S., SMALLEY, R. L. and BEAMS, H. W. 1961. Cytological studies on the intestinal epithelial cells of *Ascaris lumbricoides suum*. *Trans. Amer. Micr. Soc.* **80**: 103–18.

73. KIKUCHI, G., RAMIREZ, J. and BARRON, E. S. G. 1959. Electron transport system in *Ascaris lumbricoides*. *Biochim. Biophys. Acta*, **36**: 335–42.

74. KMETEC, E. and BUEDING, E. 1961. Succinic and reduced diphosphopyridine nucleotide oxidase systems of *Ascaris* muscle. *J. biol. Chem.* **236**: 584–91.

75. KMETEC, E., MILLER, J. H. and SWARTZWELDER, J. C. 1962. Isolation and structure of mitochondria from *Ascaris lumbricoides* muscle. *Exp. Parasit.* **12**: 184–91.

76. KROTOV, A. I. 1957. (Content of acetylcholine-like substances and cholinesterase in *Ascaris* tissues.) (In Russian.) *Byull. Eksp. Biol. Med. Moscow*, **43**: 95–7.

77. KRUSBERG, L. R. 1960. Hydrolytic and respiratory enzymes of species of *Ditylenchus* and *Pratylenchus*. *Phytopathology*, **50**: 9–22.

78. LAPAGE, G. 1935. The second ecdysis of infective nematode larvae. *Parasitology*, **27**: 186–206.

*79. LAPAGE, G. 1937. *Nematodes parasitic in animals*. Methuen & Co. Ltd., London. 172 pp.

80. LEE, D. L. 1958. Digestion in *Leidynema appendiculata* (Leidy, 1850), a nematode parasitic in cockroaches. *Parasitology*, **48**: 437–47.

81. —— 1960. The effect of changes in the osmotic pressure upon *Hammerschmidtiella diesingi* (Hammerschmidt, 1838) with reference to the survival of the nematode during moulting of the cockroach. *Parasitology*, **50**: 241–6.

82. —— 1962. The distribution of esterase enzymes in *Ascaris lumbricoides*. *Parasitology*, **52**: 241–60.

83. —— Unpublished data.

84. LEWERT, R. M. and LEE, C. L. 1957. The collagenaselike enzymes of skin-penetrating helminths. *Amer. J. Trop. Med. Hyg.* **6**: 473–7.

85. LINCICOME, D. K. 1953. A streptococcal decapsulation test for detection of hyaluronidase activity in animal parasites. *Exp. Parasit.* **2**: 333–40.

86. MACKIN, J. G. 1936. Studies on the morphology and life history of nematodes in the genus *Spironoura*. *Illinois Biol. Monographs*, **14**: 7–64.

87. MAGGENTI, A. R. 1962. The production of the gelatinous matrix and its taxonomic significance in *Tylenchulus* (Nematoda: Tylenchulinae). *Proc. helm. Soc. Wash.* **29**: 139–44.

88. MAKIDONO, J. 1956. Observations on *Ascaris* during fluoroscopy. *Amer. J. Trop. Med. Hyg.* **5**: 699–702.

89. MASSEY, V. and ROGERS, W. P. 1950. The intermediary metabolism of nematode parasites. I. The general reactions of the tricarboxylic acid cycle. *Aust. J. Sci. Res.* B, **3**: 251–64.

90. MELLANBY, H. 1955. The identification and estimation of acetylcholine in three parasitic nematodes (*Ascaris lumbricoides*, *Litomosoides carinii* and the microfilariae of *Dirofilaria repens*). *Parasitology*, **45**: 287–94.

91. MONNÉ, L. 1955. On the histochemical properties of the egg envelopes and external cuticles of some parasitic nematodes. *Ark. Zool.* **9**: 93–113.

92. MOORE, H. B. 1931. The muds of the Clyde Sea area. III. Chemical and physical conditions, rate and nature of sedimentation, and fauna. *J. Mar. biol. Ass.* **17**: 325–58.

93. MORGAN, G. T. and McALLEN, J. W. 1962. Hydrolytic enzymes in plant-parasitic nematodes. *Nematologica*, **8**: 209–15.

94. MUELLER, J. F. 1929. Studies on the microscopical anatomy and physiology of *Ascaris lumbricoides* and *Ascaris megalocephala*. *Z. Zellforsch.* **8**: 361–403.

95. MYUGE, S. G. 1957. (On the physiological specificity of the bulb nematode, *Ditylenchus allii* Beij.) (In Russian.) *Zoologischeski Zhurnal*, **36**: 620–2.

96. —— 1959. (Development of parasitism in plant nematodes.) (In Russian.) *Helminthologia, Bratislava*, **1**: 43–50.

97. NICHOLAS, W. L. 1959. The cultural and nutritional requirements of free-living nematodes of the genus *Rhabditis* and related genera. 161–8. In: *Plant Nematology. Min. Ag. Fish. Food, Tech. Bull.* **7**: H.M. Stat. Off., London.

98. NIMMO-SMITH, R. H. (Personal communication.)

99. NIMMO-SMITH, R. H. and KEELING, J. E. D. 1960. Some hydrolytic enzymes of the parasitic nematode *Trichuris muris*. *Exp. Parasit.* **10**: 337–55.

100. NORTON, S. and DE BEER, E. J. 1957. Investigations on the action of piperazine on *Ascaris lumbricoides*. *Amer. J. Trop. Med. Hyg.* **6**: 898–905.

101. OSCHE, G. 1952. Die Bedeutung der Osmoregulation und des Winkverhaltens für freilebende Nematoden. *Z. Morph. Ökol. Tiere*, **41**: 54–77.

102. OVERGAARD NEILSON, C. 1949. Studies on the soil microfauna. II. The soil inhabiting nematodes. *Natura Jutlandica*, **2**: 1–131.

*103. OYA, H., COSTELLO, L. C. and SMITH, W. N. 1963. The comparative biochemistry of developing *Ascaris* eggs. II. Changes in cytochrome c oxidase activity during embryonation. *J. cell. comp. Physiol.* **62**: 287–93.

104. PANNIKAR, N. K. and SPROSTON, N. G. 1941. Osmotic relations of some metazoan parasites. *Parasitology*, **33**: 214–23.

105. PIRIE, N. W. 1960. Report of the biochemistry department: eelworm hatching factors, 112–3 (A. J. Clarke). *Rothamsted Experimental Station Report for* 1960.

106. POLLACK, J. K. 1957. The metabolism of *Ascaris lumbricoides* ovaries. 3. The synthesis of alanine from pyruvate and ammonia. *Aust. J. Biol. Sci.* **10**: 465–74.

107. POLYAKOVA, O. I. 1959. (Hyaluronidase or ' penetrating factor ' in *Dictyocaulus filaria*.) (In Russian.) *Helminthologia, Bratislava*, **1**: 281–5.

*108. PROSSER, C. L. and BROWN, F. A. 1961. *Comparative animal physiology*. 2nd Ed., 688 pp. W. B. Saunders Co., Philadelphia.

109. RATHBONE, L. 1955. Oxidative metabolism in *Ascaris lumbricoides* from the pig. *Biochem. J.* **61**: 574–79.

*110. READ, C. P. 1960. The carbohydrate metabolism of worms. 3–34. In: *Comparative physiology of carbohydrate metabolism in Heterothermic animals*. Univ. of Washington Press, Seattle.

111. REDMOND, J. R. 1955. The respiratory function of hemocyanin in Crustacea. *J. cell. comp. Physiol.* **46**: 209–47.

112. RHOADES, H. L. and LINFORD, M. B. 1961. A study of the parasitic habit of *Paratylenchus projictus* and *P. dianthus*. *Proc. helm. Soc. Wash.* **28**: 185–90.

113. ROBERTS, L. S. and FAIRBAIRN, D. 1963. Metabolism of *Nippostrongylus brasiliensis* (Nematoda: Trichostrongyloidea). *J. Parasit.* **49** (Suppl.): 51.

114. ROBINSON, J. 1962. *Pilobolus* spp. and the translation of the infective larvae of *Dictyocaulus viviparus* from faeces to pasture. *Nature, Lond.* **193**: 353–4.

115. ROCHE, M. and TORRES, C. M. 1960. A method for *in vitro* study of hookworm activity. *Exp. Parasit.* **9**: 250–6.

116. ROGERS, W. P. 1940. The effects of environmental conditions on the accessibility of third stage Trichostrongyle larvae to grazing animals. *Parasitology*, **32**: 208–25.

*117.   1941. Digestion in parasitic nematodes. III. The digestion of proteins. *J. Helminth.* **19**: 47–58.

118.   1949. The biological significance of haemoglobin in nematode parasites. I. The characteristics of the purified pigments. *Aust. J. Sci. Res.* **B, 2**: 287–303.

119.   1949. The biological significance of haemoglobin in nematode parasites. II. The properties of the haemoglobins as studied in living parasites. *Aust. J. Sci. Res.* **B, 2**: 399–407.

120.   1952. Nitrogen catabolism in nematode parasites. *Aust. J. Sci. Res.* **B, 5**: 210–22.

*121.   1960. The physiology of infective processes of nematode parasites; the stimulus from the animal host. *Proc. roy. Soc.* **B, 152**: 367–86.

*122.   1962. *The nature of parasitism. The relationship of some metazoan parasites to their hosts.* Academic Press Inc., N.Y. 287 pp.

123. ROGERS, W. P. and LAZARUS, M. 1949. Glycolysis and related phosphorus metabolism in parasitic worms. *Parasitology*, **39**: 302–14.

*124. ROGERS, W. P. and SOMMERVILLE, R. I. 1963. The infective stage of nematode parasites and its significance in parasitism. *Advances in parasitology*, **1**: 109–77.

125. ROHDE, R. A. 1960. Acetylcholinesterase in plant-parasitic nematodes and an anticholinesterase from asparagus. *Proc. helm. Soc. Wash.* **27**: 121–3.

126. RONALD, K. 1963. The effects of physical stimuli on the larval stage of *Terranova decipens*. III. Electromagnetic spectrum and galvanotaxis. *Canad. J. Zool.* **41**: 197–217.

127. ROTHSTEIN, M. 1963. Nematode biochemistry—III. Excretion products. *Comp. Biochem. Physiol.* **9**: 51–9.

128. ROTHSTEIN, M. and TOMLINSON, G. A. 1961. Biosynthesis of amino acids by the nematode *Caenorhabditis briggsae*. *Biochim. Biophys. Acta*, **49**: 625–7.
129. SAVEL, J. 1955. Études sur la constitution et le metabolisme proteiques d'*Ascaris lumbricoides* Linné, 1758. *Rev. Path. comp.* **55**: 52–121; 213–79.
130. SAZ, H. J. and VIDRINE, A. 1959. The mechanism of formation of succinate and proprionate by *Ascaris lumbricoides* muscle. *J. biol. Chem.* **234**: 2001–5.
*131. SAZ, H. J. and WEIL, A. 1962. Pathway of formation of α-methylvalerate by *Ascaris lumbricoides*. *J. biol. Chem.* **237**: 2053–6.
132. SCHOLANDER, P. F. 1960. Oxygen transport through hemoglobin solutions. *Science*, **131**: 585–90.
133. VON SCHULZ, E. 1931. Betrachtungen über die Augen freilebender Nematoden. *Zool. Anz.* **95**: 241–4.
134. SCHWABE, C. W. 1957. Observations on the respiration of free-living and parasitic *Nippostrongylus muris* larvae. *Amer. J. Hyg.* **65**: 325–37.
135. SCHWARTZ, B. 1921. Effects of secretions of certain parasitic nematodes on coagulation of the blood. *J. Parasit.* **7**: 144–50.
*136. SHEPHERD, A. M. 1962. *The emergence of larvae from cysts in the genus* Heterodera. Comm. agric. Bur., Farnham Royal, England. 90 pp.
137. SILVERMAN, P. H. 1963. Exsheathment mechanisms of some nematode infective larvae. *J. Parasit.* **49** (Suppl.): 50.
138. SIMMONDS, R. A. 1958. Studies on the sheath of fourth stage larvae of the nematode parasite *Nippostrongylus muris*. *Exp. Parasit.* **7**: 14–22.
139. SMITH, M. H. 1963. Some aspects of the combination of *Ascaris* haemoglobins with oxygen and carbon dioxide. *Biochim. Biophys. Acta*, **71**: 370–6.
*140. SMITH, M. H. and LEE, D. L. 1963. Metabolism of haemoglobin and haematin compounds in *Ascaris lumbricoides*. *Proc. roy. Soc.* B, **157**: 234–57.
141. SOMMERVILLE, R. I. 1960. The growth of *Cooperia curticei* (Giles, 1892), a nematode parasite of sheep. *Parasitology*, **50**: 261–7.
142. STAUFFER, H. 1924. Die Lokomotion der Nematoden. *Zool. Jb.* **49**: 1–118.
143. STEINER, G. and ALBIN, F. M. 1933. On the morphology of *Deontostoma californicum* n. sp. (Leptosomatinae, Nematodes). *J. Wash. Acad. Sci.* **23**: 25–30.
144. STEPHENSON, W. 1942. The effect of variations in osmotic pressure upon a free-living soil nematode. *Parasitology*, **34**: 253–65.

145. TAYLOR, D. P. 1962. Effect of temperature on hatching of *Aphelenchus avenae* eggs. *Proc. helm. Soc. Wash.* **29**: 52–4.

146. THORNE, G. 1930. Predacious nemas of the genus *Nygolaimus* and a new genus, *Sectonema*. *J. agric. Res.* **41**: 445–66.

147. THORSON, R. E. 1953. Studies on the mechanism of immunity in the rat to the nematode, *Nippostrongylus muris*. *Amer. J. Hyg.* **58**: 1–15.

148. —— 1956. Proteolytic activity in extracts of the esophagus of adults of *Ancylostoma caninum* and the effect of immune serum on this activity. *J. Parasit.* **42**: 21–5.

149. —— 1956. The effect of extracts of the amphidial glands, excretory glands, and esophagus of adults of *Ancylostoma caninum* on the coagulation of the dog's blood. *J. Parasit.* **42**: 26–30.

150. TRACEY, M. V. 1958. Cellulase and chitinase in plant nematodes. *Nematologica*, **3**: 179–83.

*151. WALLACE, H. R. 1961. The bionomics of the free-living stages of zoo-parasitic and phyto-parasitic nematodes—a critical survey. *Helminth. Abstr.* **30**: 1–22.

*152. —— 1963. *The biology of plant parasitic nematodes.* Edward Arnold (Publishers) Ltd., London. 280 pp.

153. WALLACE, H. R. and DONCASTER, C. C. 1964. A comparative study of the movement of some microphagous, plant-parasitic and animal-parasitic nematodes. *Parasitology*, **54**, 313-26.

154. WALLACE, H. R. and GREET, D. N. 1964. Observations on the taxonomy and biology of *Tylenchorhynchus macrurus* (Goodey, 1932) Filipjev, 1936 and *Tylenchorhynchus icarus* sp. nov. *Parasitology*, **54**, 129-44.

*155. WEINSTEIN, P. P. 1960. Excretory mechanisms and excretory products of nematodes: an appraisal. 65–92. In: *Host influence on parasite physiology.* Editor: Stauber, L. A. Rutgers Univ. Press, New Brunswick.

156. WELCH, H. E. 1959. Taxonomy, life cycle, development, and habits of two new species of Allantonematidae (Nematoda) parasitic in drosophilid flies. *Parasitology*, **49**: 83–103.

157. WILSON, P. A. G. 1958. The effect of weak electrolyte solutions on the hatching rate of the eggs of *Trichostrongylus retortae-formis* (Zeder) and its interpretation in terms of a proposed hatching mechanism of strongyloid eggs. *J. exp. Biol.* **35**: 584–601.

*158. WINSLOW, R. D. 1960. Some aspects of the ecology of free-living and plant-parasitic nematodes. In: *Nematology. Fundamentals and recent advances with emphasis on plant parasitic and soil forms.* Editors: Sasser, J. N. and Jenkins, W. R. The University of North Carolina Press, Chapel Hill. 480 pp.

159. WRIGHT, K. A.   1963.   Cytology of the bacillary bands of the nematode *Capillaria hepatica* (Bancroft, 1893).   *J. Morph.* **112**: 233–59.

160.   1963.   The cytology of the intestine of the parasitic nematode *Capillaria hepatica* (Bancroft, 1893).   *J. Ultrastructure Res.* **9**: 143–55.

161. YAMAO, Y.   1951.   (Histochemical studies on endoparasites.   I. Distribution of acidic and alkaline glycerophosphatases in the intestinal cells of *Ascaris lumbricoides* L.)   (In Japanese.) *Zool. Mag., Tokyo*, **60**: 101–5.

Index

Bold type indicates reference to an illustration